CROWN ETHERS AND CRYPTANDS

Monographs in Supramolecular Chemistry

Series Editor: J. Fraser Stoddart, *University of Birmingham, U.K.*

This series has been designed to reveal the challenges, rewards, fascination, and excitement in this new branch of molecular science to a wide audience and to popularize it among the scientific community at large.

No. 1 Calixarenes
By C. David Gutsche, Washington University, St. Louis, U.S.A.

No. 2 Cyclophanes
By François Diederich, University of California, U.S.A.

No. 3 Crown Ethers and Cryptands
By George W. Gokel, Univeristy of Miami, U.S.A.

Forthcoming Title
Cyclodextrins
By J. Fraser Stoddart, University of Birmingham, U.K. and R. Zarzycki, University of Sheffield, U.K.

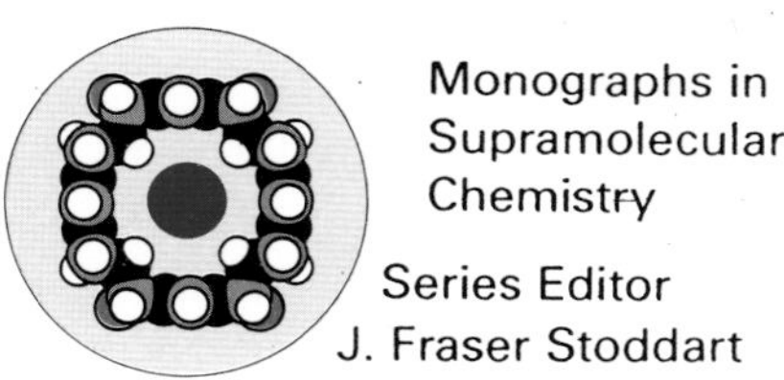

Monographs in Supramolecular Chemistry

Series Editor
J. Fraser Stoddart

Crown Ethers and Cryptands

George W. Gokel
University of Miami
Coral Gables, Florida, U.S.A.

ISBN 0-85186-996-3

A catalogue record of this book is available from the British Library

Published by The Royal Society of Chemistry,
Thomas Graham House, Science Park, Cambridge CB4 4WF

Typeset by AsTec, Newport, Saffron Walden, Essex and printed by Black Bear Press Ltd., Cambridge, England

Preface

When Professor Stoddart asked that I write a monograph on crowns and cryptands aimed at senior undergraduate and graduate students, I was pleased and honored. As I thought more about the task, it became a daunting prospect. Not quite ten years ago, Korzeniowski and I wrote a monograph solely concerned with crown and cryptand syntheses and even then it was more than 400 pages long. The number of papers in this field has burgeoned since then and encyclopedic coverage is essentially impossible.

In this book, I have attempted to give the beginner or generalist some perspective on the field and its early development. I was privileged to be part of this and can therefore relate some of it from personal experience. I have then tried to give a notion of syntheses, binding phenomena, complex structures, and to offer a representative sample of applications. So many fine scientists have contributed to this field that even some very important papers have either been left out or not given sufficient coverage. I regret this and hope that my colleagues in this field will recognize that not everything could be mentioned. I also must admit that the coverage does, to some extent, reflect my own prejudices but I hope this will not prove to be a disadvantage. In many cases, there is no reference to the original literature but the reader is directed to review articles or monographs. It is hoped that these reviews will provide the broader coverage the reader requires.

I wish to thank Ms. Leslie Brereton who helped with the manuscript and Isa Delgado, Jeanette Hernandez, Julio Medina, and Professor Angel Kaifer who offered helpful suggestions and assisted in other important ways. I also thank my co-workers generally for tolerating all the time I was away from the laboratory in pursuing this effort. In the end, I thank Fraser Stoddart for asking me to do it as it has caused me to think about the field in a way I had not previously done. I hope this volume proves helpful to those who read it.

George W. Gokel
Coral Gables, Florida
September 1990

CPK atomic model of dibenzo-18-crown-6 in the binding conformation in the absence and presence of K^+

Contents

Chapter 1 Introduction to Macrocycle Chemistry 1

1.1 Introduction 1
1.2 Crown Ether Precursor Work 2
1.2.1 Lüttringhaus' Macrocycles 2
1.2.2 Cyclo-oligomerization of Ethylene Oxide 3
1.2.3 The Furan–Acetone Tetramer 3
1.3 Pedersen's Discovery of Crown Ethers 4
1.4 Simmons' In–Out Bicyclic Amines 9
1.5 Lehn's Cryptands 10
1.6 Cram and Carbanions 12
1.7 Classes of Crown Ethers and Cryptands 14
1.8 Naturally Occurring Relatives of Crown Ethers 17
1.9 Nomenclature and its Problems 19
1.10 Toxicity of Crown Ether Compounds 21

Chapter 2 Syntheses of Crowns, Cryptands, and their Relatives 22

2.1 Introduction 22
2.2 Pedersen's First Crowns 22
2.3 The Ethyleneoxy Structural Unit 23
2.4 The Template Effect 27
2.5 Early Syntheses of Crowns 31
2.5.1 Dicyclohexano-18-crown-6 31
2.5.2 18-Crown-6 31
2.5.3 [2.2.2]-Cryptand 32
2.6 Syntheses of Crown Ethers 33
2.6.1 Monomers, Dimers, and Trimers 33
2.6.2 Nucleophiles and Bases 34
2.6.3 Leaving Groups 35
2.6.4 Solvents 38
2.6.5 High Dilution 39
2.6.6 General Conditions 40

2.7 Incorporation of Aromatic Subunits 40
2.8 Nitrogen-containing Macrocycles 42
2.8.1 Protection of Nitrogen 43
2.8.2 Lactam Formation Followed by Reduction 46
2.8.3 Direct Incorporation of Nitrogen 49
2.8.4 Syntheses of Lariat Ethers 49
2.8.5 Multi-armed Lariat Ethers 51
2.8.6 Heteroaromatic Subunits 53
2.9 Crown Ethers Containing Sulfur 56
2.10 Syntheses of Cryptands 58
2.11 Spherands, Cavitands, and Carcerands 60
2.12 Summary 63

Chapter 3 Complexation by Crowns and Cryptands 64

3.1 Introduction 64
3.2 The Extraction Technique 66
3.3 Homogeneous Cation Binding Constants 71
3.3.1 Techniques for Determining Stability Constants 71
3.3.2 Variables Affecting Homogeneous Stability Constants 73
3.3.3 The Macrocyclic Effect 76
3.3.4 Enthalpy–Entropy Compensation 76
3.3.5 Brief Survey of Cation Complexation Constants 78
3.4 Binding Dynamics 79
3.5 Cation Transport 81
3.6 Complexation of Organic Cations 83
3.6.1 Ammonium Cations 83
3.6.2 Chiral Recognition 86
3.6.3 Arenediazonium Cations 87
3.6.4 Hydronium Cation 89
3.6.5 Co-operative Systems 91
3.6.6 A Polymeric Ionophore 92
3.7 Molecular Complexation 92
3.8 Anions 95

Chapter 4 Structural Aspects of Crowns, Cryptands, and their Complexes 99

4.1 Introduction 99
4.2 Uncomplexed Crown and Cryptand Ligands 99
4.3 Simple Crown Ether Complexes of Metal Cations 103

4.3.1 Crown Complexes of Cations that Are 'Too Large' 106
4.3.2 Crown Complexes of Cations that Are 'Too Small' 109
4.3.3 Complexes of Functional Crown Ether Derivatives 110
4.3.4 Lariat Ether Complexes 111
4.3.5 Bibracchial Lariat Ether Complexes 113
4.4 Complexes of Cryptands 115
4.4.1 Simple Cryptand (Cryptate) Complexes 115
4.4.2 Ditopic Cryptands 118
4.4.3 Inclusion *vs.* Exclusion Complexes 119
4.5 Crown Ether Complexes of Other Ions and Molecules 120
4.6 Podand Complexes 126

Chapter 5 Applications of Crowns and Cryptands 129

5.1 Introduction 129
5.2 Solubilization Phenomena 130
5.2.1 Solubilization of Salt 132
5.2.2 Crown Activated Reagents 133
5.3 Cation Deactivation 136
5.3.1 Deactivation of Lithium Aluminum Hydride ($LiA1H_4$) 136
5.3.2 Deactivation of the Cannizzarro Reaction 136
5.4 Anion Activation: Phase Transfer Catalysis 137
5.4.1 Principles of Phase Transfer Catalysis 138
5.4.2 Displacement Reactions 141
5.4.3 Superoxide Chemistry 142
5.4.4 Aromatic Substitution Reactions 142
5.4.5 Other Examples 144
5.4.6 Comparison of Crowns and Quaternary 'Onium Salts 144
5.5 Sensors and Switching 145
5.5.1 Switching Modes 145
5.5.2 Ionization Control of pH-Switching 145
5.5.3 Photochemical Control 150
5.5.4 Thermal Switching 151
5.5.5 Redox Switching 151
5.6 Polymeric Crown Ethers 156
5.7 Membranes and Channels 158
5.7.1 Cation Transport 158

5.7.2 Cation-conducting Channels 160
5.7.3 Hydrophobic Macrocycles: Micelles and Membranes 162
5.8 Chirality, Complexation, and Enzyme Models 164
5.8.1 Cram's Chiral Crown Ethers 164
5.8.2 Rebek's Allosteric System 167
5.8.3 Cram's Protease Model 167
5.8.4 Receptor Molecules 169
5.9 Miscellaneous Applications and Developments 173
5.9.1 Metal-templated Catenane Formation 173
5.9.2 π-Complexed Catenane Formation 175
5.10 The Future of Crown Ether Chemistry 177

Chapter 6 Additional Reading 178

6.1 Books on Crown Ethers, Cryptands, and Polyethers 178
6.2 Monographs on Phase Transfer Catalysis 179
6.3 Reviews and Articles 179

Subject Index 186

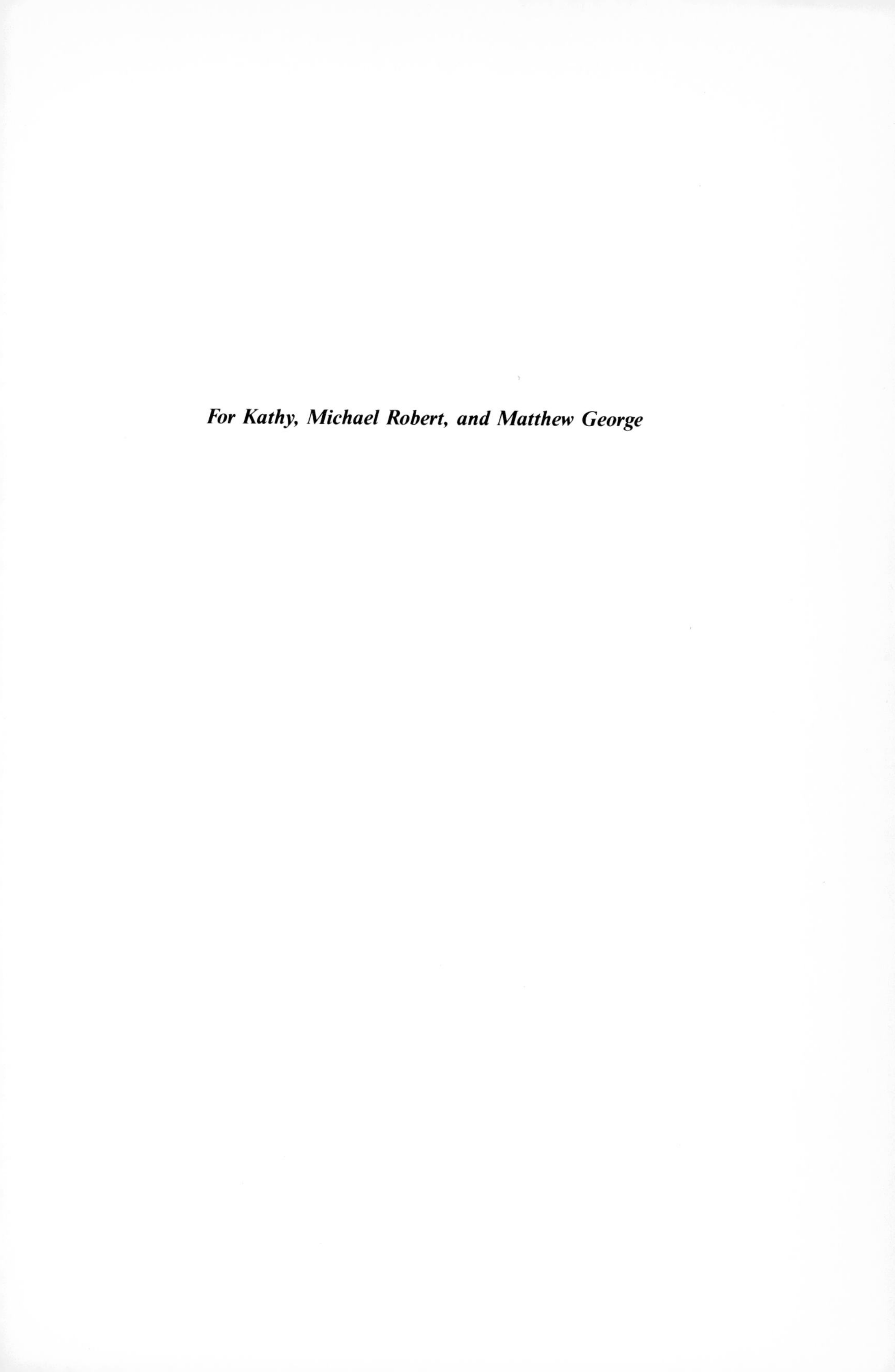

For Kathy, Michael Robert, and Matthew George

CHAPTER 1

Introduction to Macrocycle Chemistry

'Synthetic organic chemists fall into two groups: those who prepare old, naturally occurring compounds and those who prepare new compounds. The synthetic targets of the former group are provided by the evolutionary chemistry of nature. The synthetic targets of the latter group are designed by the investigator'

D.J. Cram in *Cyclophanes*, eds. P.M. Keehn and S.M. Rosenfield, Academic Press, New York, 1983, Vol. 1, p. 14.

1.1 Introduction

The award of the 1987 Nobel Prize in chemistry to Charles Pedersen, Donald Cram, and Jean-Marie Lehn, marks an important turning point in the history of organic chemistry. For most of the 150 years between Wohler's synthesis of urea and the present time, the focus of organic chemists has been on molecules containing only covalently linked atoms. This Nobel Prize delineates that time in history when the focus changed from covalent to non-covalent interactions. Of course, this focus could only change as a result of the extensive, important, and often brilliant efforts of innumerable organic chemists to understand structures, shapes, and reaction mechanisms.

During the post-war period of this century, we may view organic chemistry as having gone through two major phases. The period from the end of World War II to the early 1970s has been regarded by many as the golden age of physical organic chemistry. During this time the basic mechanistic principles that control all organic reactions came into sharp focus for the first time.

In the late 1960s, concurrent with the award of the Nobel Prizes to such giants of synthesis as Woodward and Barton, chemistry entered the golden age of natural product synthesis. This era was brought not to a culmination but certainly to a pinnacle in the remarkable synthesis by the groups of Woodward and Echenmoser of vitamin B_{12}. At about the time that the fields of total synthesis and synthetic methods were in full force and vigor, a new area of chemistry was in its infancy. This new field was based on historical developments, as are all innovations, but it would find a new direction for science in terms of approach and thinking. One can always find related phenomena nearby any major accomplishment. It is not our intention to down-play any of the important, if somewhat peripheral, contributions but rather to focus attention on those which history now identifies as the major contributions. There is, however, some important background that should be mentioned before we consider Pedersen's discovery of the crown ethers.

1.2 Crown Ether Precursor Work

1.2.1 Lüttringhaus' Macrocycles

It is often the case that chemistry assumes a national character. It is the nature of scientific training that individuals new to the field learn from those more senior. Within any given department, university, or country, traditions are passed down in science as they are in other facets of life. The Germans were important and early players in both macromolecular chemistry and the synthesis of large ring compounds. The Author has heard the comment that the Germans have a 'fascination for ring systems'. It is fortunate for the rest of the scientific community that the Germans have this fascination since many important advances arose from it. This fascination is completely justified, especially since Kekulé recognized the ring structure of benzene and in so doing made one of the most fundamental advances organic chemistry has ever known.

Lüttringhaus was interested in preparing large ring molecules which might concatenate. His hope was to prepare structures which would interlock and have at least interesting and perhaps unusual properties. One of the substrates he chose for his studies was the resorcinol molecule (1,3-dihydroxybenzene). He reacted this nucleophile with substituted diol derivatives. The result was that he isolated several structures of the type shown in **1**.

Unfortunately, Lüttringhaus' compounds lacked sufficient donor groups to exhibit cation binding properties. Although the synthesis was interesting and the products were fascinating, the remarkable aspect of crown ethers and their relatives would eventually prove to be their ability

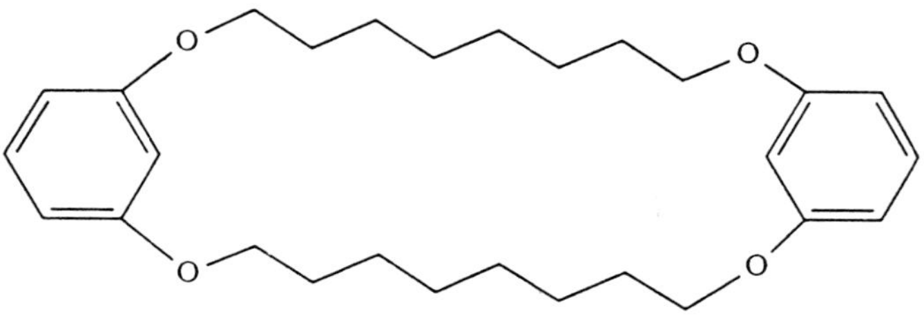

O O O O

1

to capture or include other species. It was Pedersen who recognized this ability. Indeed, Pedersen recognized this possibility even before he synthesized the first crown.

1.2.2 Cyclo-oligomerization of Ethylene Oxide

In the mid-1950s, Stewart, Waddan, and Borrows obtained a British Patent on a process for the cyclo-oligomerization of ethylene oxide. Treatment of oxirane (**2**) with alkyl aluminum, zinc, and magnesium compounds produced dioxane and other cyclic materials. Principal among the macrocyclic oligomers was the cyclic tetramer of ethylene oxide, 'EO-4' (**3**). These workers recognized that the macrocycle had interesting properties but they failed to appreciate its potential. They noted that it should prove useful as a high temperature solvent but failed to carry the effort farther. Some years later, Dale showed that this synthetic approach was more general (see Chapter 2).

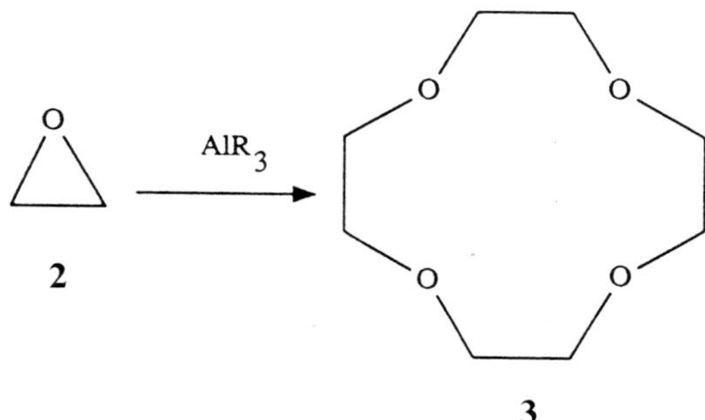

1.2.3 The Furan–Acetone Tetramer

When a protonic or Lewis acidic catalyst is present in solution with furan and acetone, cyclic products are formed. Of these, the cyclic tetramer is best known and well characterized (see Chapter 2). The formation of this compound is important because it represents a system in which there is a relatively high level of organization and the four donor atoms are focused to the center of the cavity. In early studies on these compounds, Ackman, Brown, and Wright called the products 'anhydrotetramers' because the analytical data corresponded to the ketone and furan starting materials minus four water molecules. A name was suggested for the cyclic tetramer of acetone and furan: 2,2,7,7,12,12,17,17-octamethyl-21,22,23,24-tetraoxaquaterene (**4**).

It should probably be noted that although these compounds bear a superficial resemblence to porphyrins, they are not fully conjugated.

CH_3 CH_3 O + O Lewis Acid

4

1.3 Pedersen's Discovery of Crown Ethers

The reader interested in the details of the crown ether discovery should consult two very interesting accounts that appeared recently. Pedersen's own account of his life and also the pathway that led him to the discovery of crowns is recounted in some detail in his Nobel Prize address. His former co-worker at Dupont, Harold Schroder, presented an account of Pedersen's life and discovery at the 1987 International Macrocycle Symposium held in Hiroshima, Japan. That account has also been published. The two views, while not identical, both offer interesting perspectives on the discovery. What follows is a much less detailed account that attempts to touch the essential features of the work.

Charles J. Pedersen was a chemist working in the Elastomer Chemicals Department of the Dupont Chemical Company's research facility in Wilmington, Delaware. This is the corporate research center for Dupont and like many such centers is sub-divided into areas of chemical interest. At the time of the discovery, there was a department at Dupont called the Central Research Department that was usually referred to as CRD. It was in this Department that Howard Simmons, whose contributions are discussed later, was working simultaneously with Pedersen.

The story of Pedersen's first crown synthesis, which has often been referred to as accidental, is by now very well known. Pedersen was interested to prepare non-cyclic phenolic derivatives which could complex or coordinate various divalent cations. In his Nobel Prize address, Pedersen indicated that he was particularly interested in the vanadyl (VO) group. Pedersen recognized that charge-dense cations require charged donor groups for them to be bound most effectively. He therefore undertook the preparation of a *bis*(phenol) derivative **5** which could be deprotonated at moderate pH.

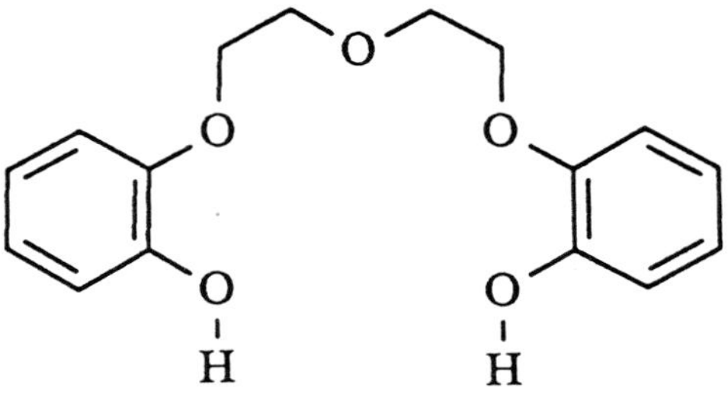

5

Pedersen described his observations concerning the synthesis as follows:

> 'As I proceeded, I knew that the partially protected catechol...was contaminated with about 10 percent unreacted catechol. But I decided to use this mixture for the second step anyway since purification would be required at the end. The reactions were carried out as outlined and gave a product mixture in the form of an unattractive goo. Initial attempts at purification gave a small quantitity (about 0.4 percent yield) of white crystals which drew attention by their silky, fibrous structure and apparent insolubility in hydroxylic solvents.
>
> The appearance of the small quantity of unknown compound should have put me in a quandary. It probably was not the target compound because that would be obtained in higher yield. My objective was to prepare and test a particular compound for a particular purpose. Had I followed this line, I would have doomed the crown ethers to oblivion until such a time as another investigator would retrace my steps and make the better choice at the critical moment. Crown ethers, however, were in no danger, because of my natural curiosity. Without hesitation, I began study of the unknown'.

Pedersen obtained the compound he desired (**5**) after removal of the protecting groups. He also characterized the small amount of neutral material that ultimately proved to be dibenzo-18-crown-6 (**6**). The synthesis of dibenzo-18-crown-6 was indeed accidental. The synthesis of a very closely related material, however, was intentional. Moreover, Pedersen was pursuing cation complexing agents when he prepared the first crowns. The synthesis of the first crown ethers, while accidental, should obviously not be represented as simple good luck.

electron rich cyclic systems. Examples of the various complexes and some of the macrocycles are shown in structures **9–16**.

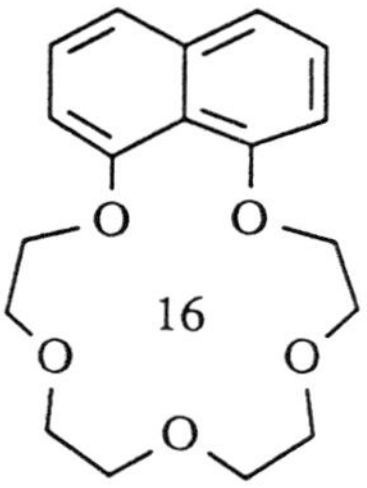

9 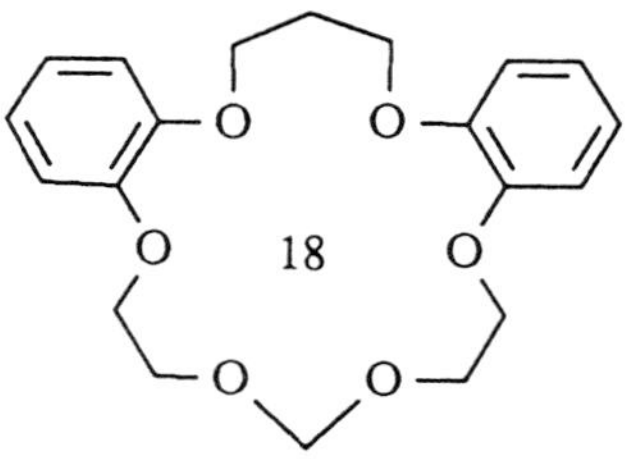 **10**

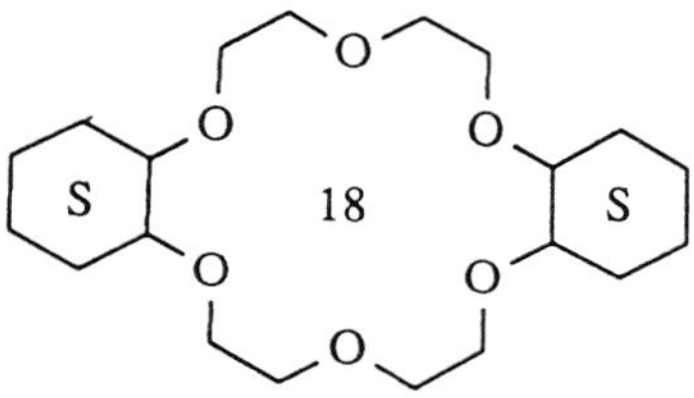

11 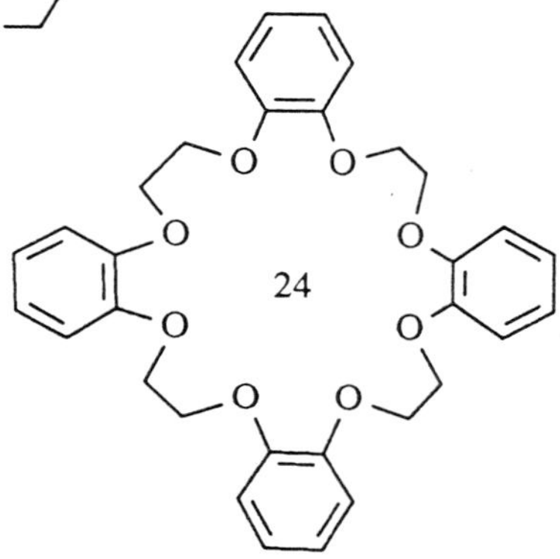 **12**

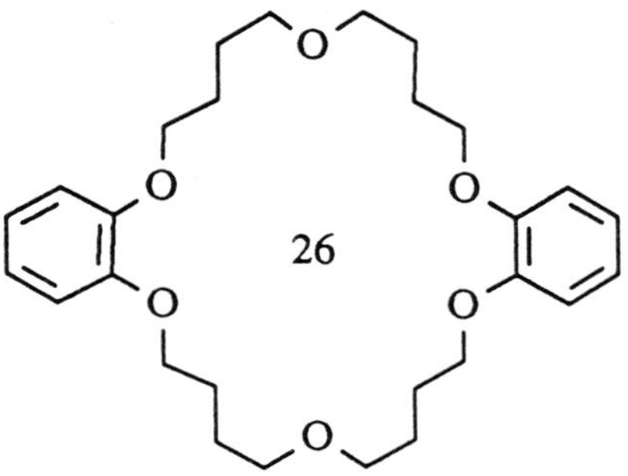

13 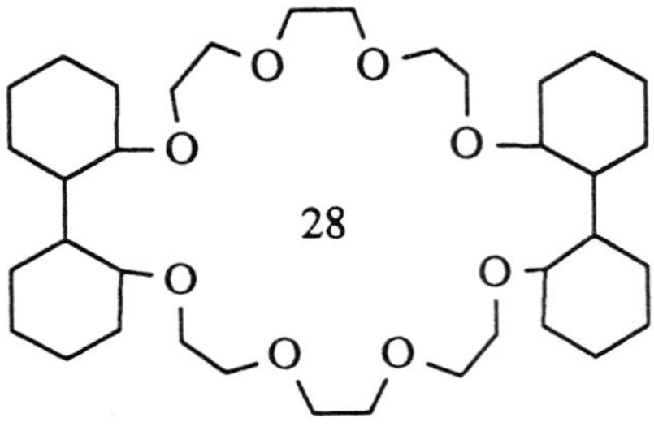 **14**

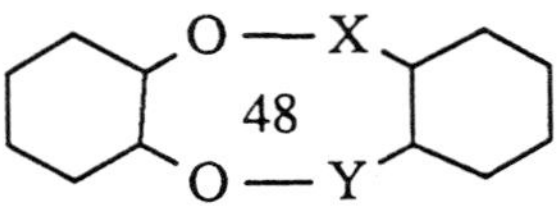

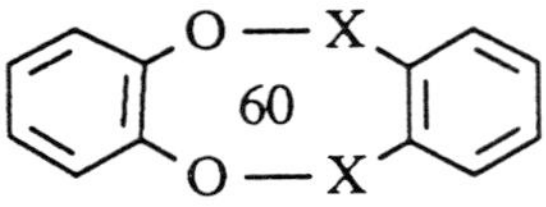

$X = (CH_2CH_2O)_9$

$Y = (CH_2CH_2O)_5$

15 **16**

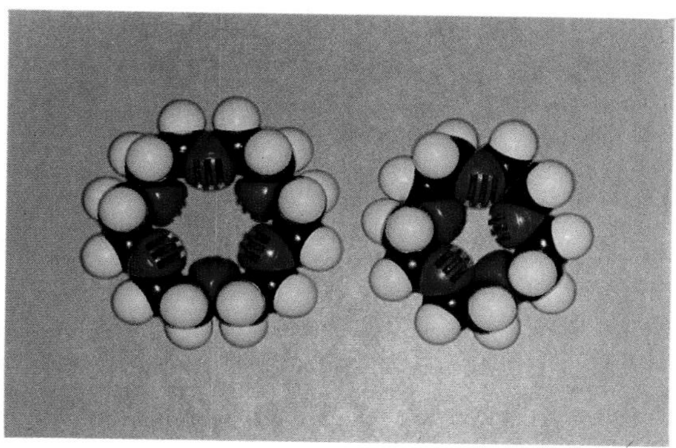

CPK atomic models of 15-crown-5 and 18-crown-6

1.4 Simmons' In–Out Bicyclic Amines

Howard Simmons, of Simmons-Smith cyclopropanation fame, was working along a somewhat different line of inquiry. Simmons has said publicly that he was inspired to attempt the synthesis of the bicyclic amines from his readings of old German literature. Perhaps he, like Lüttringhaus, fell prey to the German fascination for rings. In any event, his notion was to prepare a system which had a three-dimensional structure. At the time of their synthesis, three-dimensional structures were known but such structures were mostly quite small, quite rigid, or both. Simmons realized that a system containing three polymethylene chains anchored at opposite ends to a nitrogen atom would have considerable conformational flexibility and therefore quite interesting properties. He directly recognized that the chains would be flexible and that the nitrogen atoms would be subject to rapid inversions. At a minimum, compounds could exist in conformations in which (i) the electron pairs of both nitrogens were pointed inward (ii) the electron pairs of both nitrogen atoms were pointed outward, and (iii) the conformation in which one electron pair was pointed inward and one was pointed outward. These intriguing substances have generally been referred to as 'in–out' bicyclic amines. The conformations identified were those called 'in–in' (**18**), 'out-out' (**17**), or 'in–out'.

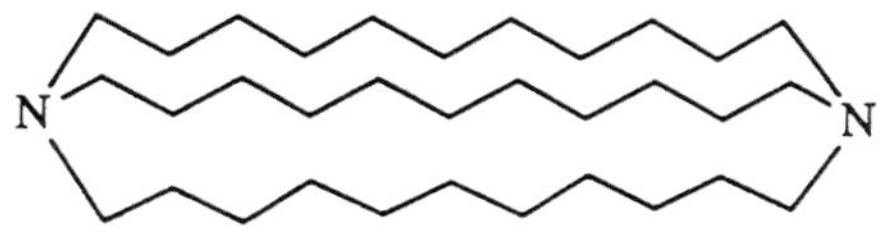

17

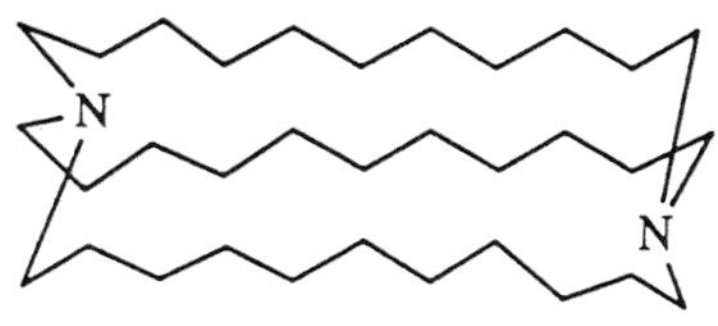

18

It appears from the subsequent literature that Simmons' interest in these compounds was as novel structures and also as systems capable of interesting conformational variation. We see below that the same basic elements were incorporated in the cryptands with the addition of donor atoms.

1.5 Lehn's Cryptands

Simultaneous with this remarkable activity at Dupont Chemical Company, a young Frenchman by the name of Jean-Marie Lehn had just returned to Strasbourg from post-doctoral studies with Professor Woodward at Harvard. Lehn considered the various topologies possible for enveloping molecular species. Of course, the crowns and the bicyclic amines represented only two of numerous possibilities. He realized that the presence of donor groups within a three-dimensional, flexible array would permit a cation not only to be complexed but to be encapsulated. Addition of a third donor group strand to the crown ethers makes these molecules the analog of a first solvation shell for a cation. The compounds that Lehn and his co-workers prepared incorporated features of both the bicyclic amines and Pedersen's crown ethers. Lehn recognized the advantage of complete encapsulation and distinguished his contribution by calling these compounds 'cryptands'. The root word crypt comes from the Latin *crypta* which, in turn, comes from the Greek *kruptos* meaning 'hidden or to hide'. The idea of this evocative name is that the cation, when bound, is completely hidden from the bulk solvent that previously stabilized it.

Lehn faced the same problem that Pedersen had in naming these compounds. Though they are structures of high symmetry, the systematic nomenclature is complex indeed. Lehn suggested a system of nomencla-

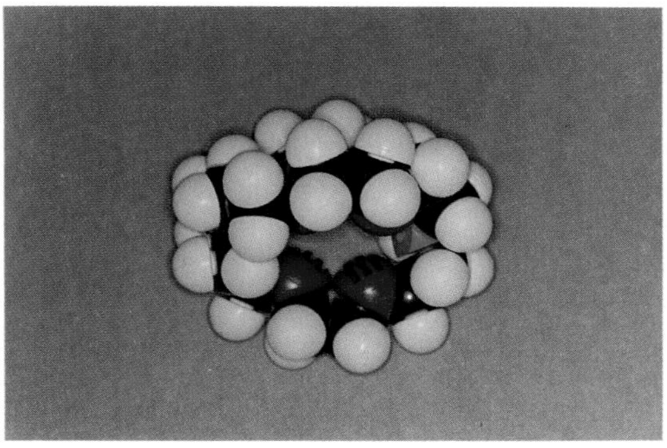

CPK atomic model of [2.2.2]-cryptand

ture based on the presumption that heteroatoms in the ring system are separated by two carbons and that three chains would be anchored at nitrogen unless otherwise specified. He thus named the cryptand comprised of three chains having two oxygen heteroatoms in each one [2.2.2]-cryptand **20** (Figure 1.1). The 2 refers not to the carbon spacing between the oxygens but to the number of oxygens in each chain. Notice that there are actually eight atoms in each chain, six carbon and two oxygen atoms.

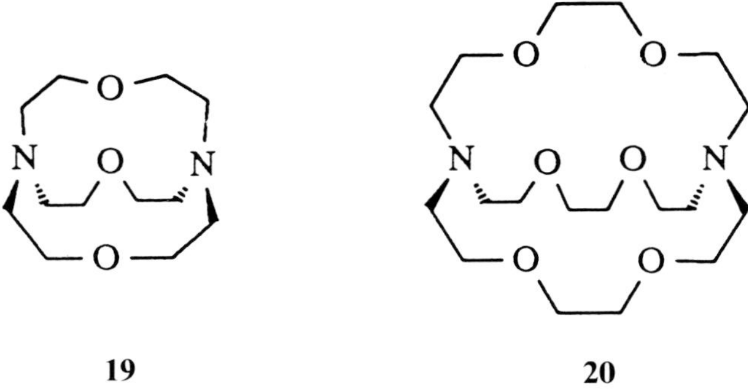

Figure 1.1 *[1.1.1]-Cryptand (***19***) and [2.2.2]-cryptand (***20***)*

It is a very interesting question to ask, from the historical perspective, whether Lehn's ideas resulted simply by combining the notions of Pedersen and Simmons. This was not the case. The author has heard the following story related concerning the invention of the cryptands. In the late 1960s, Lehn and his co-worker Jean-Pierre Sauvage, had prepared the first example of [2.2.2]-cryptand (**20**) and already knew some of the compound's properties at a time when, coincidentally, Simmons visited Strasbourg. There was, according to the story, some discussion between Lehn and Sauvage about the extent of revelations for patent reasons that should be made to Simmons during his visit. At that time, the cryptands were still unpublished but both concept and synthesis were well in hand. As befits a great and confident scientist, the decision was made to reveal the cryptands in full, should the subject arise. It is often said that great minds think alike, and Simmons had, by that time, conceived structures similar to the cryptand. When he mentioned these to Lehn, Lehn produced a vial containing the very compound Simmons had conceived but not yet brought to hand.

As was the case with Pedersen, Lehn immediately began to explore the incorporation of other heteroatoms, the preparation of molecules having different ring sizes, and the complexation of a variety of species. As early as the late 1960s, he disclosed materials having oxygen, sulfur and nitrogen heteroatoms within the structures and also reported the complexation of cations by these materials. Figure 1.2 shows [2.1.1]-cryptand (**21**),

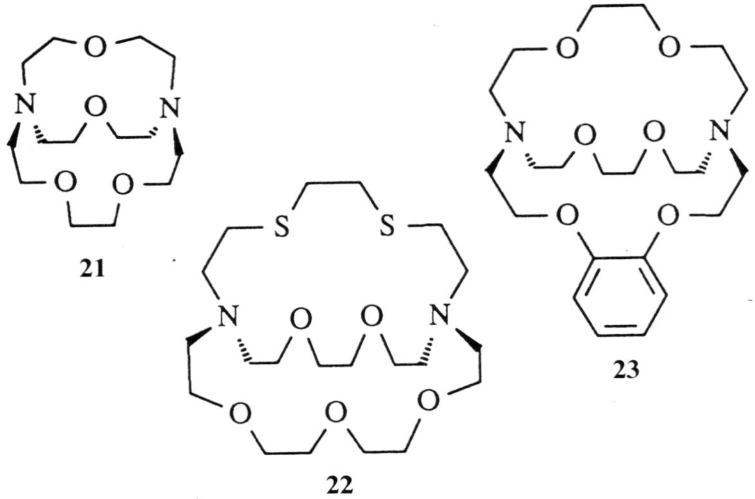

Figure 1.2 *[2.1.1]-Cryptand (**21**), [3.2.2]-cryptand (**22**), and benzo[2.2.2]-cryptand (**23**)*

[3.2.2]-cryptand containing two sulfur atoms in a single bridge (**22**), and benzo[2.2.2]-cryptand (**23**).

Lehn has suggested a modification to his abbreviated nomenclature to accommodate a variety of substituents and heteroatoms in cryptands. Rather than discuss it, we will exemplify by naming the central and right compounds in Figure 1.2. The central compound would be referred to as [3.2.2_S]-cryptand (**22**) and the right hand structure is [2.2.2_B]-cryptand (**23**). The advantages and disadvantages of this approach will be obvious to the reader.

1.6 Cram and Carbanions

Donald J. Cram, the founding father of carbanion chemistry, working at the University of California in Los Angeles, undertook extensive investigations that led, in part, to the publication in 1965 of his textbook titled *Fundamentals of Carbanion Chemistry*. Cram recognized that carbanions exist in various aggregation states and realized that their chemical activity could be increased by separating the countercation from the anion. One possible way to do this would be to fully solvate the cation using a cyclic polyether. Cyclic tetramers (**4**) were known (see above) that derived from furan and acetone. This macrocycle, Cram reasoned, might complex a lithium cation and thereby activate (help desolvate) the associated anion.

In the mid-1960s, Cram asked one of his associates to look into this possibility. Unfortunately, for whatever reason this line of inquiry failed to bear fruit. When the papers of Pedersen began to appear in 1967, Cram realized that his original idea, although quite important, had been too

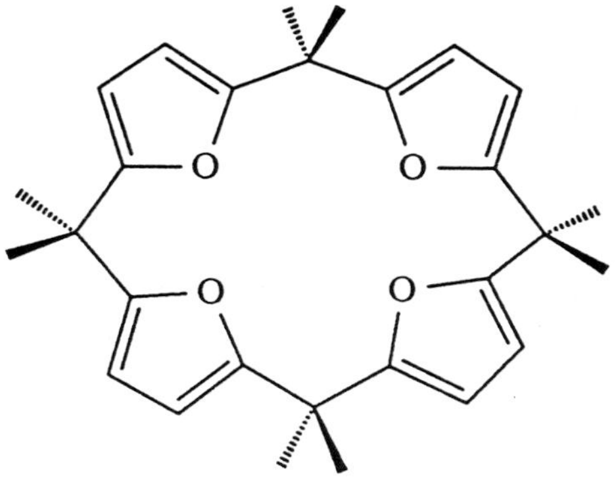

4

narrow in scope. Cram's thinking expanded quickly from the original idea of using crown ethers or structures related to them as anion activators. Eventually, he focused his ideas on the preparation of chiral macrocycles which could permit the stereochemical differentiation of optically active primary ammonium salts.

The Author, as a graduate student, attended a seminar given by Cram who characterized his vision as follows. He conceived of a beaker containing a racemic ammonium salt. This aqueous solution would be poured into a separatory funnel. A chloroform solution of a chiral crown ether would be added and the two solutions shaken together. The crown ether, because of appropriate chiral barriers integral to its structure, would selectively complex and extract only one enantiomer. The two solutions would then be separated in the normal fashion. One enantiomer would partition into the crown-containing chloroform solution and one would remain in the aqueous phase. To the Author's knowledge, this single step extraction was never realized but it has certainly been approached in a variety of ways. Indeed, Cram and his co-workers studied the resolution of primary alkylamines and amino acids. The type of crown ethers studied most extensively in those early days are illustrated in the Figure 1.3. How such ligands may be used for enantiomeric discrimination is discussed in Chapter 5.

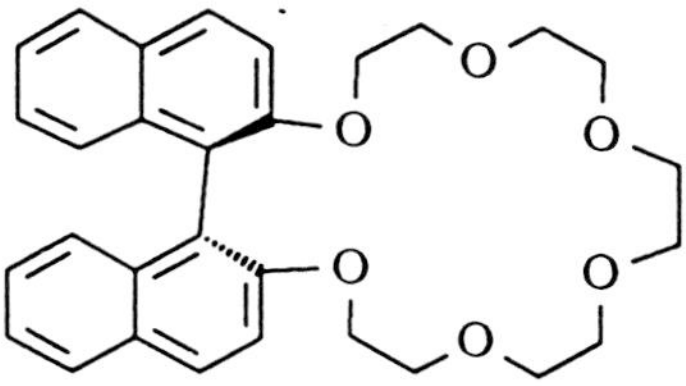

24 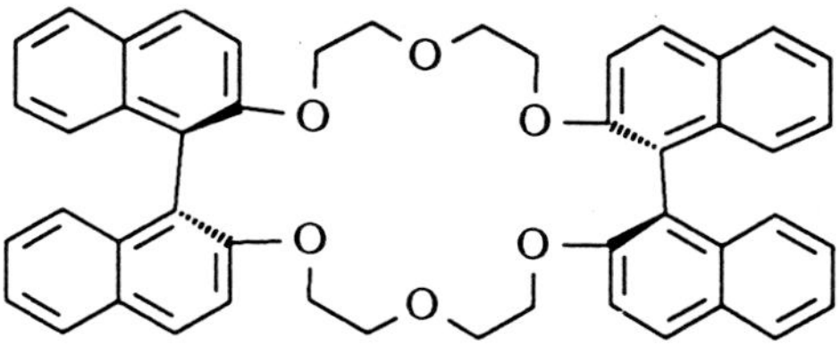 25

Figure 1.3 *Optically active binaphtho-20-crown-6 (**24**) and an isomer of* bis-*(binaphtho)-22-crown-6 (**25**)*

1.7 Classes of Crown Ethers and Cryptands

The family of macrocycles has now grown so broad that it is impossible to characterize its full extent in a single section of such a short monograph. It is possible only to give a general notion of the types of compounds that have been prepared. The crown ethers are macrocyclic polyether compounds generally, but not always, comprised of oxygen, nitrogen, and/or sulfur heteroatoms (usually) separated by two carbon atoms or their equivalent. Probably the only definition of crown ether that remains adequate today considering the huge number of structures and structural variations extant, is that they are macrocyclic systems containing at least one macrocyclic ring (macroring). In order to be a crown ether, a compound must be cyclic and must contain donor groups such as oxygen. Replacement of oxygen by sulfur, nitrogen, and other heteroatoms has been explored extensively and has led to azacrown ethers, thiacrown ethers, and other derivatives. A few examples are shown in structures **26—31**.

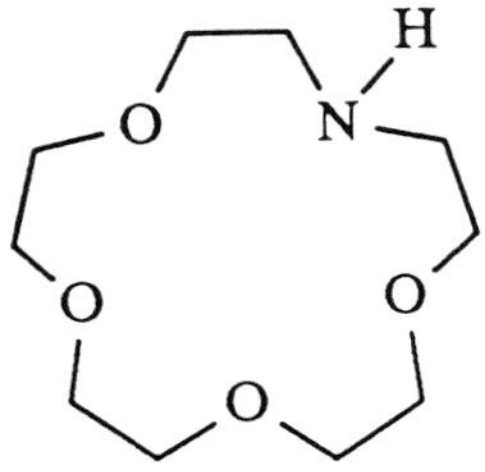

26

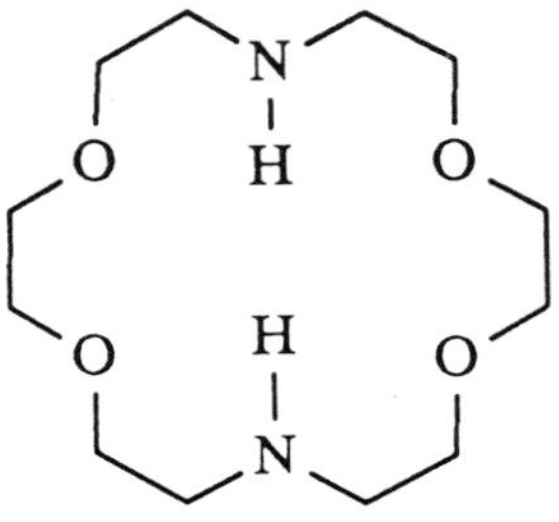

27

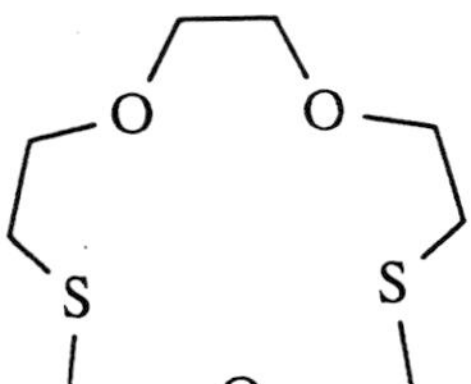

28

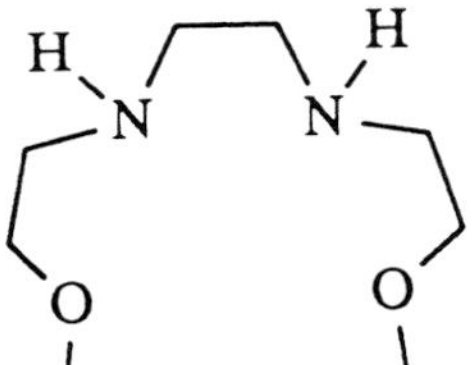

29

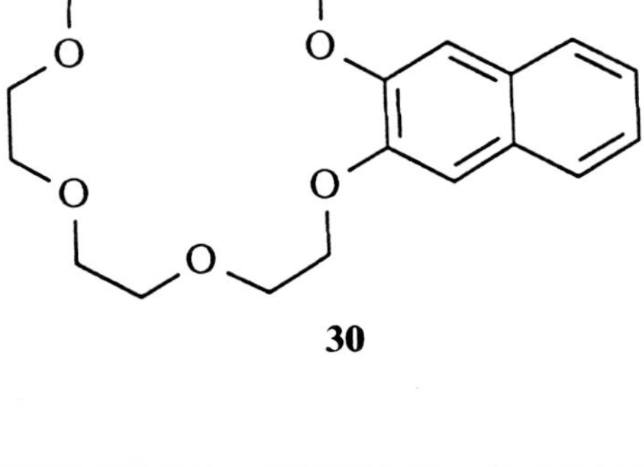

30

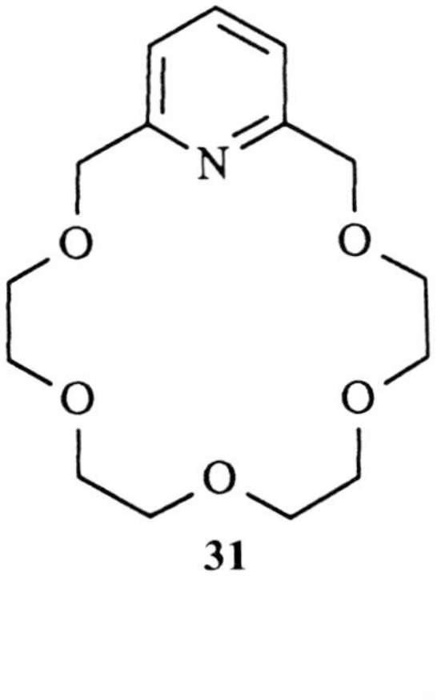

31

The cryptands form a general class of compounds which are obviously related to crown ethers. The cryptands represent the first important example of a three-dimensional cation binder and it is clear from his 1973 article in *Structure and Bonding* that Lehn conceived of a much wider variety of topologies than just the cryptands. Many of these topologies have now been realized. The compounds having more carbon chains and donor groups than were found in the original cryptands have been prepared and many quite different structures having high rigidity are also known. It should also be noted that cryptands of great structural variety have been prepared. The structures of a few are shown (**32**—**36**).

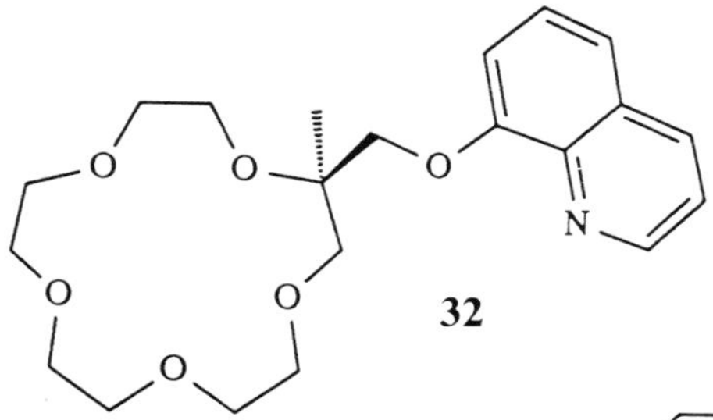

32

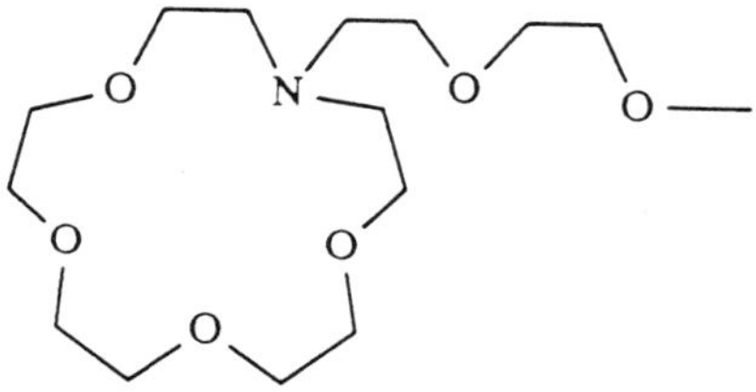

33

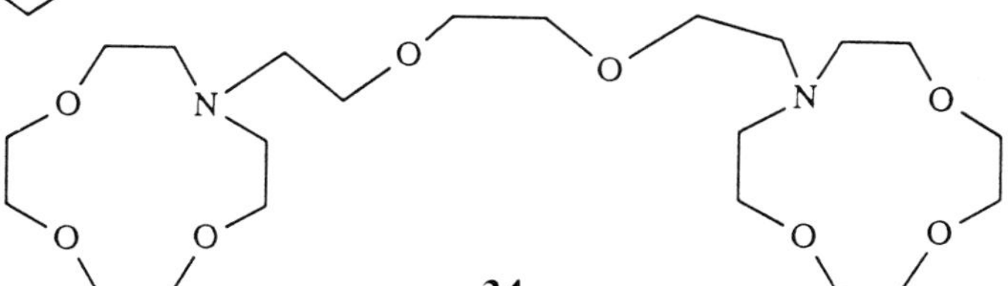

34

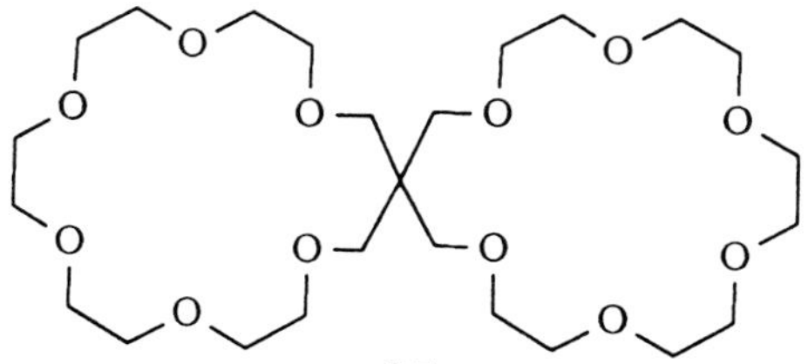

35

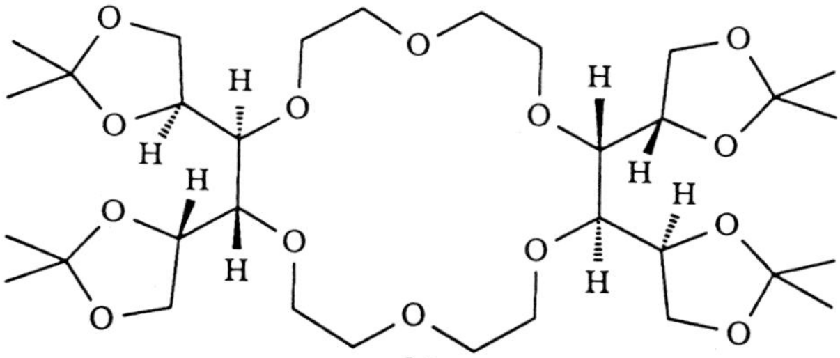

36

The three-dimensional, enveloping ligands first adumbrated by Lehn have now led to classes such as the spherands, the cavitands, and the lariat ethers, as well as numerous other materials. In elaborating these topologies, scientists have had various goals in mind. The development of spherands for example, was prompted by a desire for tight fit, clearly oriented donor groups, and three-dimensionality. The design principles for each group of structures is obviously different and one should not consider the ligands in terms of better or worse. Each has a different set of properties and therefore advantages and difficulties. As with any major area of chemistry, the ideas and inspirations diverge then merge and diverge again. A bicyclic cryptand and two spherands are included among the four structures shown (**41—43**). The structure having the generally pyramidal appearance (**40**) for example, was prepared by Schmidtchen as a receptor and would not be appropriate for cation complexation. The compounds shown are: **41**, a spherand; **42**, a bicyclic cryptand; **43**, an aromatic spherand (R is CH_3O); and **44**, an anthraquinone-cryptand.

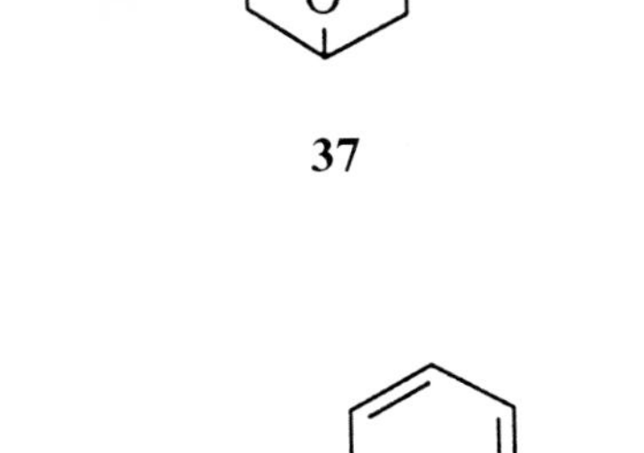

37

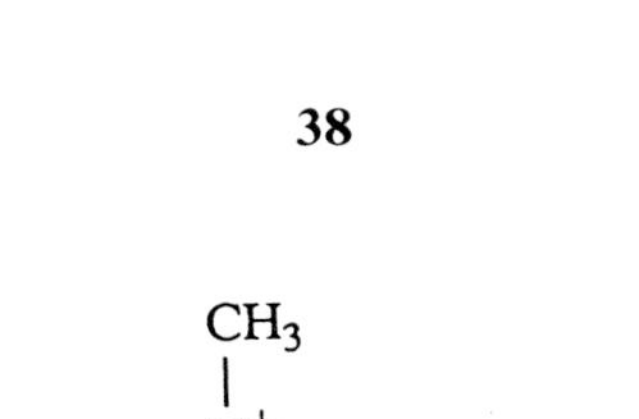

38

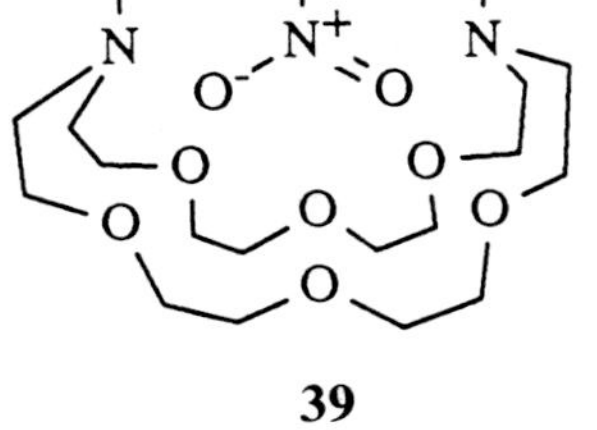

$X = (CH_2)_n$

39

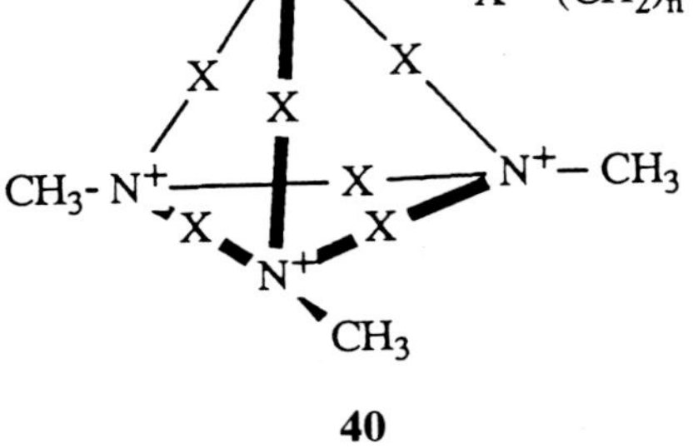

40

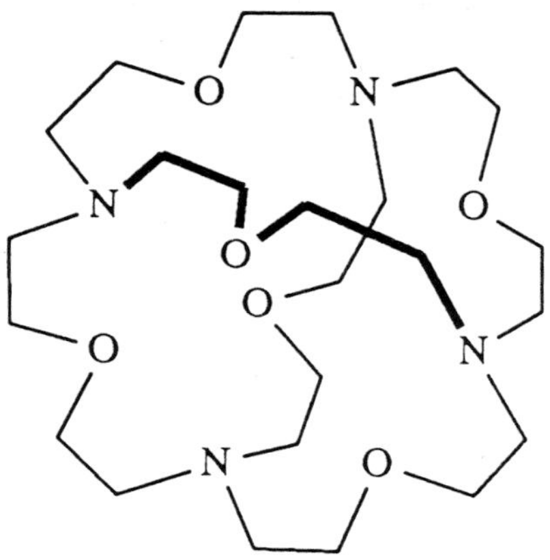

41

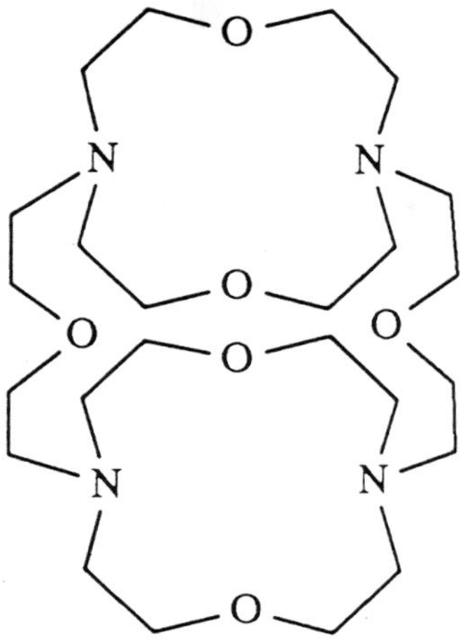

42

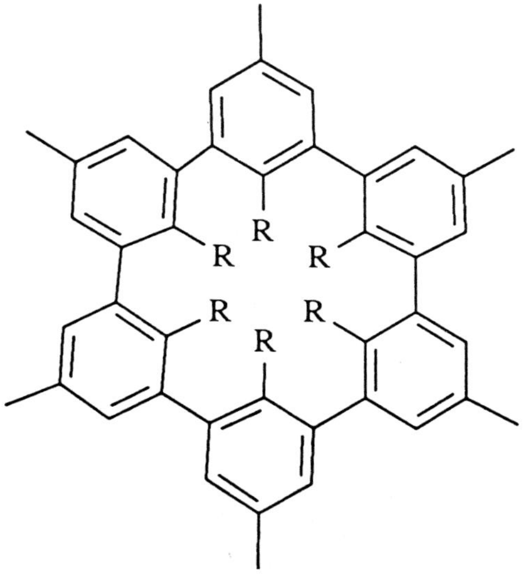

43

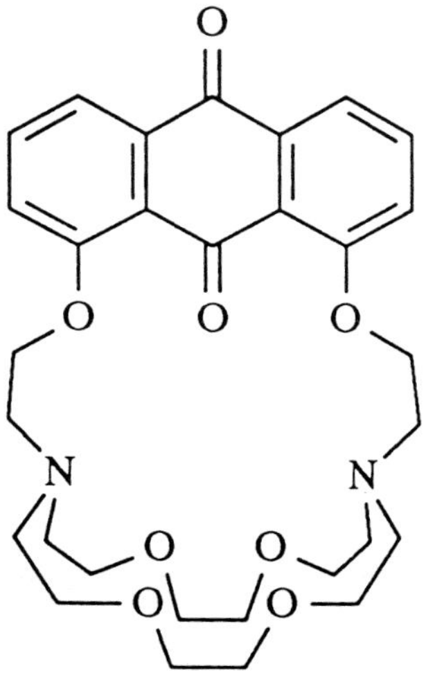

44

1.8 Naturally Occurring Relatives of Crown Ethers

Considering that Nature prepared structures capable of cation binding eons before the thought ever occurred to man, it is probably arrogant to consider the following compounds as analogs of crown ethers. Alas, it is a peculiar idiosyncrasy of human thought, to organize Nature in terms of human concepts. The development of crown ethers and cryptands was as important in terms of concept and direction as it was in terms of the huge amount of practical chemistry that resulted. Once the concepts were recognized and understood, the remarkable properties of many natural systems could better be understood. Compounds such as valinomycin (Figure 1.4), a compound referred to as a 'transport antibiotic,' could be understood in terms of the principles discovered during crown and cryptand research.

Nature has found ways to complex cations using substances such as valinomycin (Figure 1.4) and nonactin (Figure 1.5) using not only simple ethers but carboxyl groups, ketones, and other residues. The compounds that have thus far inspired the greatest interest to chemists working in the macrocycle area are those capable of binding cations. Compounds that can bind and transport cations are referred to as *ionophores*.

Figure 1.4 *Valinomycin*

CPK atomic model of the naturally-occurring ionophore valinomycin showing the oxygen-lined, K^+-binding cavity

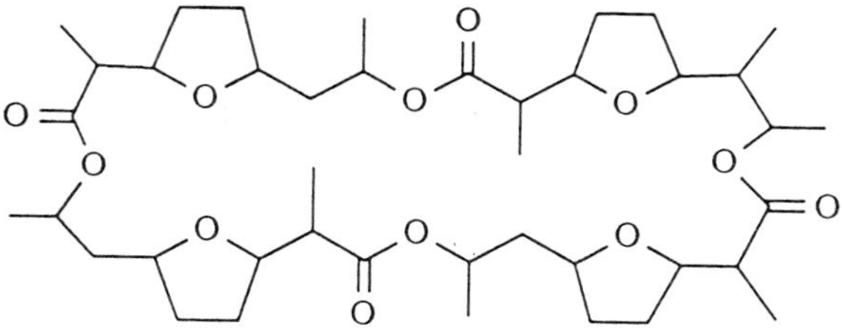

Figure 1.5 *Nonactin*

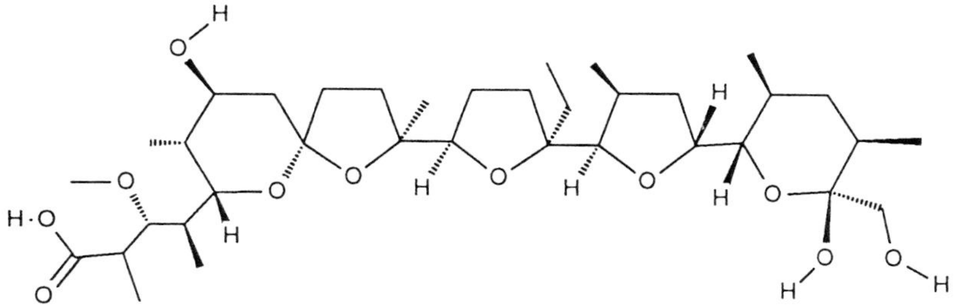

Figure 1.6 *Monensin*

Monensin (Figure 1.6) is an acyclic ionophore that cyclizes using the carboxyl group at one end and a hydroxyl group at the other. This results in a pseudo-cyclic cation binder.

1.9 Nomenclature and its Problems

From the very beginning, Pedersen recognized that systematic nomenclature was absurdly complicated for routine discussion of such complex molecules. The shorthand method he suggested has proved both useful and versatile. Likewise, the semi-systematic, shorthand nomenclature developed by Lehn for the cryptands has proved serviceable and extensible. As a practical matter, however, more complex structures have defied simple nomenclature. Indeed, a form of hieroglyphics or picture writing has been required in order for chemists to discuss structures intelligently. Of course, structural drawings have been central to organic chemistry for at least a hundred years but simple names are always needed for conversation. Some attempts have been made to offer simplified but quasi-systematic nomenclature.

Vögtle and Weber recognized this problem and offered a system to distinguish ligand types and complexes. The nomenclature system builds on Lehn's terminology. Lehn called his three-dimensional ligands crypt*ands* and their complexes crypt*ates*. Vögtle and Weber suggested that crown ethers be called *coronands* and their complexes *coronates*. A compound that is a crown ether in all respects except that it is non-cyclic may be called an opened-chain crown ether but this is cumbersome terminology. Instead, the term *podand* was suggested and complexes of such structures are therefore *podates*. This nomenclature system has been extended as shown by Table 1.1.

The cryptand–cryptate nomenclature suggested by Lehn is extensively used. The extended rules have been adopted for podands, spherands, cavitands, and carcerands. The system has proved less popular for naming crown ethers although it is used in some cases. The lack of popularity for crown ethers may be due to the fact that many of the simplest crown structures were already known before this system was suggested. As new structural types evolved, they did not necessarily fit the 'coronand' designation. Nevertheless, the term coronand is a good one to use in discussion of generic molecular species.

It is probably worth brief mention that some compounds have been given names by their inventors that are still widely used despite their inappropriateness or incorrectness. A good example is dicyclohexano-18-crown-6. Pedersen first prepared it as a mixture of isomers by hydrogenation of dibenzo-18-crown-6 (**6**) and called it 'dicyclohexyl-18-crown-6'. The preferred name is 'dicyclohexano' in analogy to 'dibenzo'.

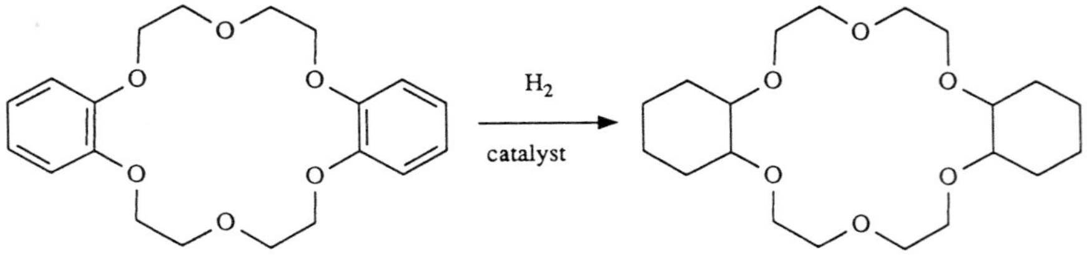

H_2, catalyst

6 **11**

One additional note is in order. Some years ago, the Author presented a lecture at a local meeting of the American Chemical Society in Washington, D.C. Robert Fox, a polymer scientist at the Naval Research Laboratory in Washington, attended the lecture and became intrigued with the problem of nomenclature. In due course, a nomenclature system was suggested that used IUPAC principles and leaned on experience with polymer systems. Using this system, 15-crown-5 would be named cyclo[pentakis-(oxyethylene)] rather than the more formal 1,4,7,10,13-pentaoxacyclopentadecane. The paper is cited among the references.

Table 1.1 *Nomenclature for Crowns, Cryptands, and their Relatives*

Free Ligand	*Complex*	*Other name for ligand*
Podand	Podate	glyme, polyethylene glycol
Coronand	Coronate	crown ether
Podando–coronands	Podando–coronates	lariat ether
Cryptand	Cryptate	
Spherand	Spherate	
Cavitand	Cavitate	
Carcerand	Carcerate	

1.10 Toxicity of Crown Ether Compounds

Some comment on the toxicity of macrocyclic polyether compounds is certainly in order. We noted in section 1.2.2 that 12-crown-4 (**3**) (or EO-4 as it was called when first prepared) was available by cyclo-oligomerization of ethylene oxide as early as the late 1950s. This reaction was of interest to chemists at Dow who undertook toxicity studies. It was discovered that the compound caused testicular atrophy when inhaled by mice at very low concentrations in the controlled atmosphere. Obviously, this was cause for concern. Moreover, Pedersen reported a deleterious effect on the eyes of a dog when dicyclohexano-18-crown-6 was applied. Although work continued on crown ethers, efforts (especially in industrial laboratories) were generally hampered by these findings. Indeed, throughout the 1970s, as crown ether chemistry gained prominence, toxicity was an oft-discussed issue at symposia and meetings.

More recently, a broader range of studies on crown compounds has been undertaken. The compounds tested were found not to be carcinogenic and the toxicity levels were, generally, low. 15-Crown-5, for example, produced redness in skin tests but its toxicity in mice was reported to be only 800 mg kg^{-1}. This means that more than 50 grams of 15-crown-5 would have to be ingested by a 150 pound human for the toxic effect to be serious. Izatt pointed out that this toxicity is about half that of aspirin, a compound not generally regarded as dangerous and indeed often ingested.

Of course, only a limited number of crown ether compounds have been tested. Some exhibit mild toxicity but some have salutary biological effects. The toxicity issue, once regarded as a major problem for crown ethers, has proved largely to be a myth. *As with all new compounds, however, caution should be exercised in handling crown ethers, cryptands, and their relatives.*

CHAPTER 2

Syntheses of Crowns, Cryptands, and their Relatives

2.1 Introduction

Considering both the number and variety of structures now falling in the general class of crown ethers and cryptands, the number of general approaches for their synthesis is remarkably small. Most approaches, as we shall see, are based on one of two synthetic strategies. The first involves simple nucleophilic substitution using one of the incipient macroring donor atoms as the nucleophile. This technique may be used for oxygen, sulfur, nitrogen, and other heteroatoms. The second general strategy is the formation of amide linkages followed by reduction to the saturated system. The latter approach is suitable primarily for azacrown and cryptand syntheses. It generally utilizes high dilution, whereas the former technique relies on what has often been called a 'template' effect. These strategies will be explored below but the synthetic story should really begin with Pedersen's seminal contribution.

2.2 Pedersen's First Crowns

Some of Pedersen's personal observations on the discovery of crowns, taken from his Nobel Prize address, may be found in section 1.3 of the previous chapter. The reader is directed to these comments if the book is being read in other than chapter order.

As luck would have it, dibenzo-18-crown-6 (**1**) was the first compound synthesized and also was one of the most interesting derivatives. It had the special properties of high crystallinity, reasonable lipophilicity, and good chemical stability. Although not the best complexing agent, it contains an eighteen membered ring having six donor groups and this has proved to be unusually effective for cation complexation.

Pedersen used the approach to dibenzo-18-crown-6 (**1**) that is shown in Scheme 2.1. In concept, the reaction is fairly straightforward. Catechol (**2**) (1,2-dihydroxybenzene) is treated with base to form the dianion. This nucleophile reacts with 2,2′-dichlorodiethyl ether (**3**) in a 'quadruple Williamson' reaction. Inexpensive sodium hydroxide could be used as the

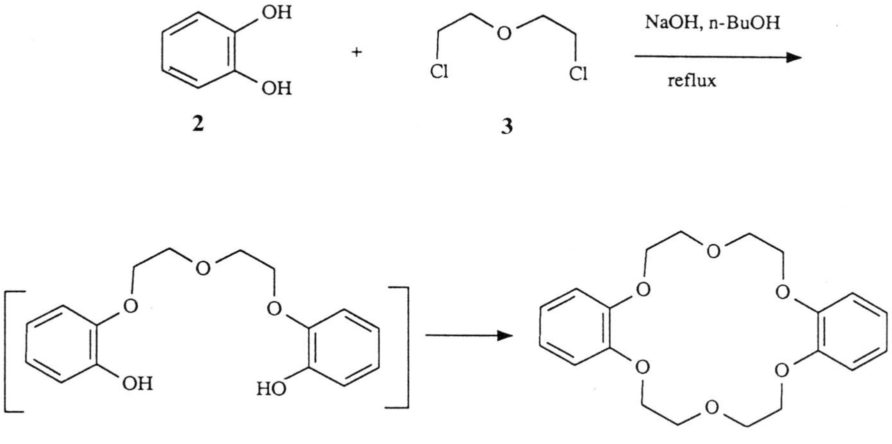

NaOH, n-BuOH

reflux

2 3 1

Scheme 2.1

base because the acid was a phenol. The choice of solvent was *n*-butanol which gave good organic solubility. Butanol is much less acidic than phenol and did not deprotonate and subsequently compete in the nucleophilic substitution reaction. The reaction was conducted at approximately 0.75 molar concentration. Pedersen did not use high dilution conditions during his synthesis of the non-cyclic complexing agents because it was not needed (see Section 2.4). The reaction was heated at the reflux temperature of *n*-butanol while stirring. Work-up involved extraction, precipitation from acetone, and recrystallization from benzene to give fine and fluffy needles in 40% yield.

This early preparation of dibenzo-18-crown-6 has been used as a model for many subsequent crown preparations. The basic technique has been widely used since and the ease of preparation has made dibenzo-18-crown-6 an important and well-studied molecule in the years since its discovery.

2.3 The Ethyleneoxy Structural Unit

It is obvious that crowns and cryptands are built largely upon the ethyleneoxy unit. This is no accident. The ethyleneoxy unit has special properties and special availability that makes it ideally suited to its application in large ring systems.

The simplest possible donor-group-containing subunit would be the methylenedioxy system, $-O-CH_2-O-$. In practice, this is the acetal of formaldehyde. It has the advantage over all other possible subunits that there are no carbon–carbon backbone interactions that could affect the macroring conformation. This unit has the disadvantage, however, that it is acid labile (Scheme 2.2).

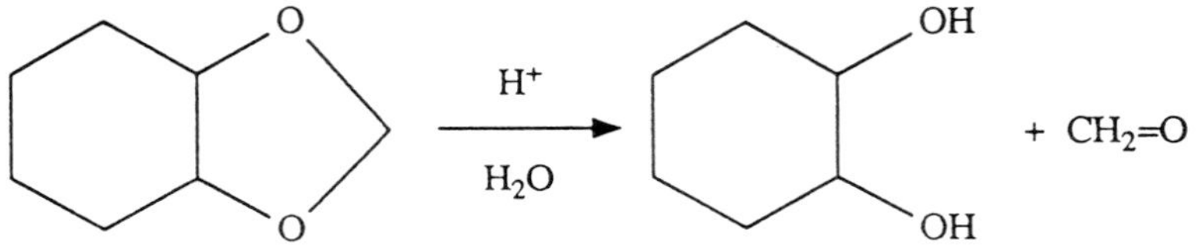

Scheme 2.2

In fact, acid lability may not be too severe a problem in the macrocycle series. It was shown a number of years ago by Gold and his co-workers that macrocyclic acetals were stabilized by cation complexation. This reduced the macrocycle's susceptibility to acid catalyzed hydrolysis, probably because complexation of the oxygen electron pairs makes them less prone to protonation, the first step in acid catalyzed hydrolysis. Even so, the number of macrocyclic acetal structures is relatively limited. Further, they appear to hold no special advantage over the historically more important ethyleneoxy systems.

Another alternative to the ethyleneoxy structural unit is the propyleneoxy unit, $-O-CH_2CH_2CH_2-O-$. The propyleneoxy unit is not subject to acid catalyzed hydrolysis but suffers from a different problem. Specifically, the propylene unit prefers to be in an extended or *anti*-arrangement. As a result, the oxygen atoms are separated by a sterically hindering $>CH_2$ unit. For the best complexation to occur, the donor groups must all be pointing inward towards the macrocycle's hole (*i.e.* the donors must be *convergent*). Unless the macrocycle adopts this conformation, solvation of the ring-bound cation will not occur and complexation will fail. The extra methylene group in 1,3-dioxypropylene makes binding more energetically costly and therefore less likely. Observed cation binding constants are generally lower for compounds that contain propyleneoxy rather than ethyleneoxy bridges. Of course,

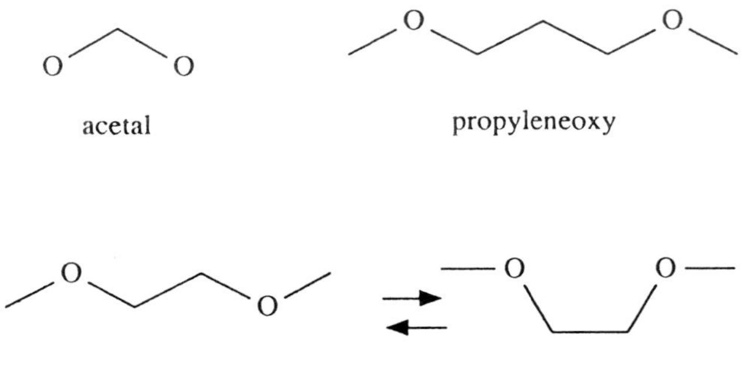

Figure 2.1 *The structural subunits of crown ethers and cryptands*

use of propyleneoxy rather than ethyleneoxy also lowers the oxygen atom density within the ring.

Compounds containing the propyleneoxy residue have been prepared and studied by the groups of Bartsch and Inoue and Hakushi. Such studies are important to confirm the presumptions outlined and always hold the possibility of revealing new insights or applications. It must be recalled that a number of macrocyclic ligands designed to bind transition metal cations contain this subunit.

In summary, then, there are three important reasons why the ethyleneoxy unit is favored over other structural units that can be imagined for macrocyclic polyethers. The first is that it is flexible; notably more flexible than the phenylenedioxy system that is found in many macrocycles. Second, the ethyleneoxy unit is relatively strain free and can easily adopt the skew or *gauche* conformation that permits easy alignment of the donor group electron pairs (Figure 2.2). The third reason that the ethyleneoxy group has proved so popular is simply a practical matter. The unit and its precursors are common industrial chemicals.

The 1,2-phenylenedioxy unit (Figure 2.2) (derived from catechol) is nearly equivalent in the structural sense to the ethyleneoxy unit in macrocycles. Its presence often facilitates synthesis since the phenolic oxygens are quite nucleophilic. Further, the aromatic ring adds rigidity to the structure and this often enhances crystallinity. Phenylenedioxy is also more hydrophobic than ethyleneoxy. The disadvantage of the phenylenedioxy group is that the two oxygen atoms are conjugated to the aromatic ring. As such, they are sp^2, rather than sp^3, hybridized. The electron pairs are therefore less available for binding to a cation situated within the ring and the electron pairs are less basic as well.

There is a fourth reason for the popularity of the ethyleneoxy group but it is historical rather than chemical. Chemistry, like most human endeavours, draws strongly on precedent. Moreover, Pedersen's choice of

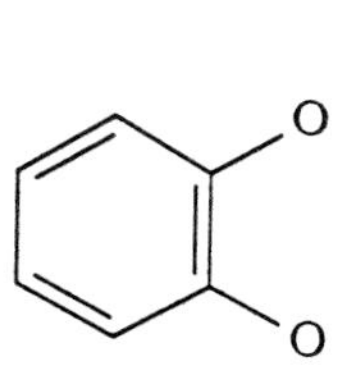

the 1,2-phenylenedioxy unit

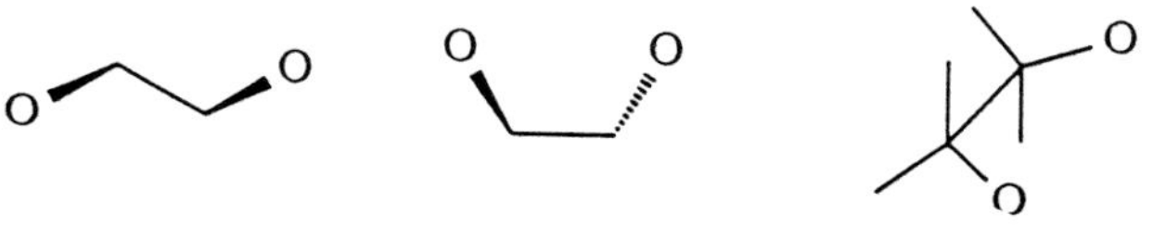

three equivalent representations of the *gauche* ethylenedioxy unit

Figure 2.2

ethyleneoxy as the basic subunit of crowns was made for good reasons and these remain valid today.

The ethyleneoxy structural unit is available from the inexpensive precursor ethylene oxide (**4**). Because ethylene oxide may readily be oligomerized or polymerized, precursor chemicals containing repeating (CH_2CH_2O) units (**5**) are also readily available and inexpensive. The variety of these two-carbon derivatives far exceeds the number of derivatives available for the one or three carbon units.

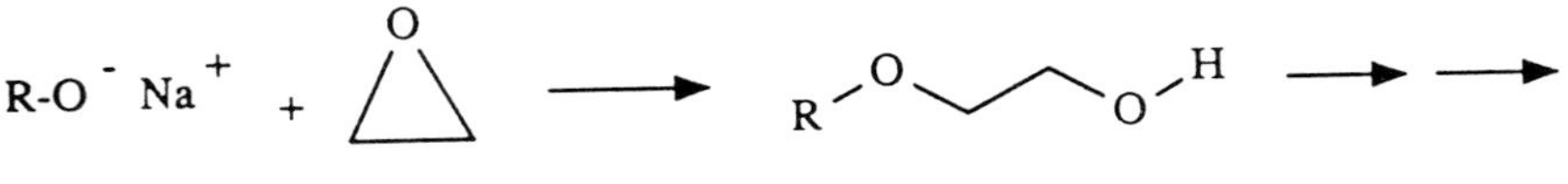

R-O⁻ Na⁺ + O → R–O–CH₂CH₂–O–H → →

4

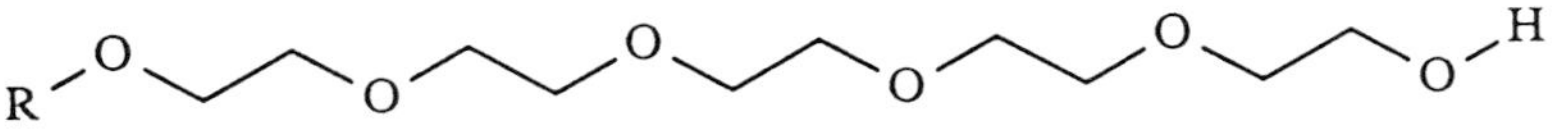

5

In principle, nearly any subunit could be included in a macrocycle. Some examples of these numerous variations are shown in structures **6–11**. Note that in most of them, the ethyleneoxy unit remains predominant. The structures include examples prepared by Pedersen, Bartsch, Inoue and Hakushi, and Cram. The non-ethyleneoxy subunit(s) in each case is highlighted for clarity.

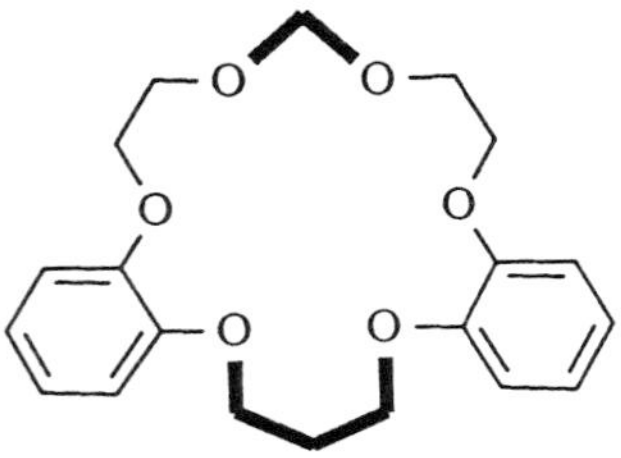

6 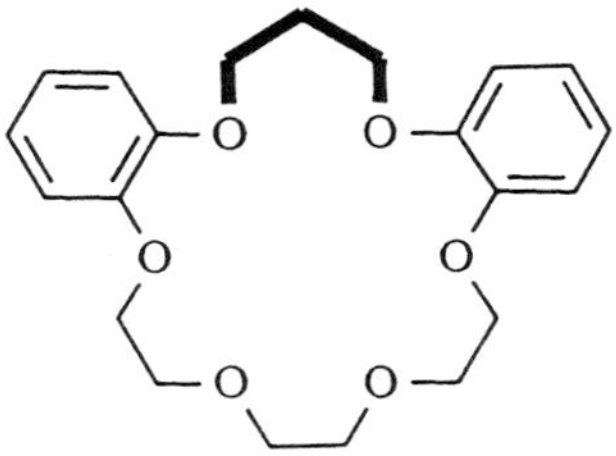 **7** 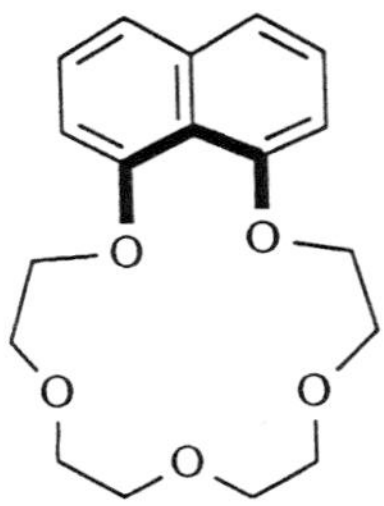 **8**

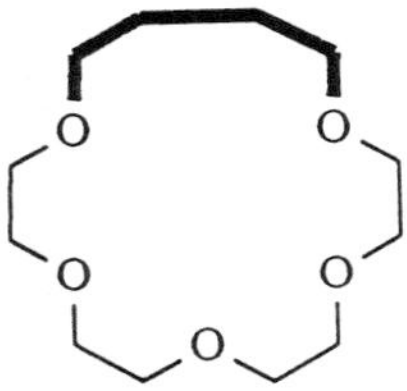

9 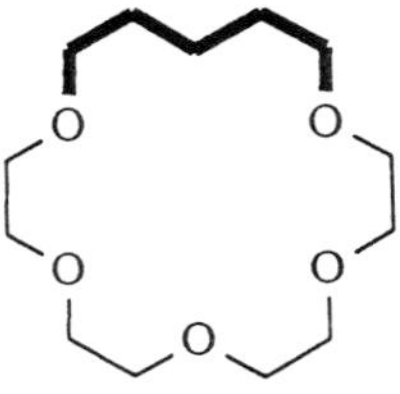 **10** 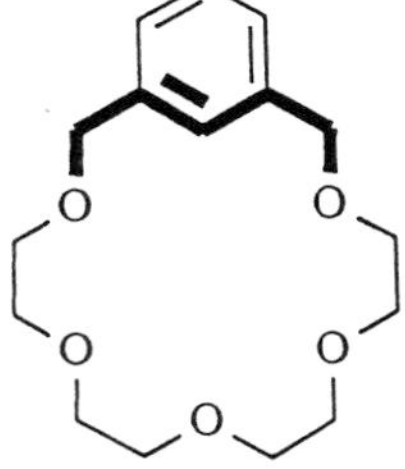 **11**

2.4 The Template Effect

Pedersen's first synthesis of dibenzo-18-crown-6, was accomplished in reasonably concentrated solution. In retrospect, there can be no doubt Pedersen recognized that the metal ion organized the transition state that led to macrocycle. This organizational principle is now known widely as 'the template effect' although Pedersen did not give it this name. It was his colleague at Dupont, Robin Greene, who first applied the term to macrocyclic polyethers. We must recall that some years earlier, Daryle Busch had shown that cyclams, the all nitrogen macrocycles, could be prepared by a template effect using transition metal ions as the organizing agents (Figure 2.3).

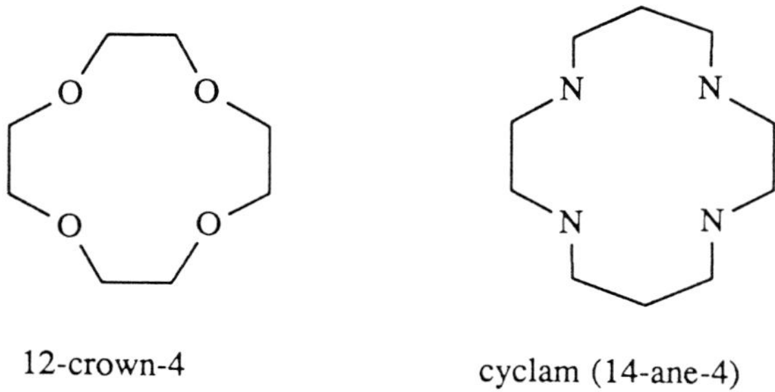

Figure 2.3 *Comparison between crown ethers and cyclams*

The principle of the template effect is straightforward. Cyclic material may be formed from a compound that is nucleophilic at one end and electrophilic at the other by reacting with itself (Path A in Figure 2.4). A second possibility exists, however: the nucleophilic or negative end of one molecule may find the electrophilic or positive end of another molecule (intermolecular path B, Figure 2.4). In the latter case, non-cyclic products (oligomers or polymers) will form.

The organic chemist is not without an arsenal of weapons in this battle. The most common tactic used to try to force the non-cyclic material to find its other end rather than another molecule is to use *high dilution conditions*. The competition between a molecule finding its other end and finding another molecule is a competition between first and second order reactions. The second order process depends upon the concentration of both materials. Thus, if we make it difficult for a molecule to find another molecule of itself, we will favor the first order, or cyclization, process. High dilution is thus used to diminish competition from the second process. The first order reaction is favored under these conditions but its rate is not enhanced.

The second approach is to enhance cyclization over oligomerization. In the case of poly(ethyleneoxy) materials, this is accomplished by the

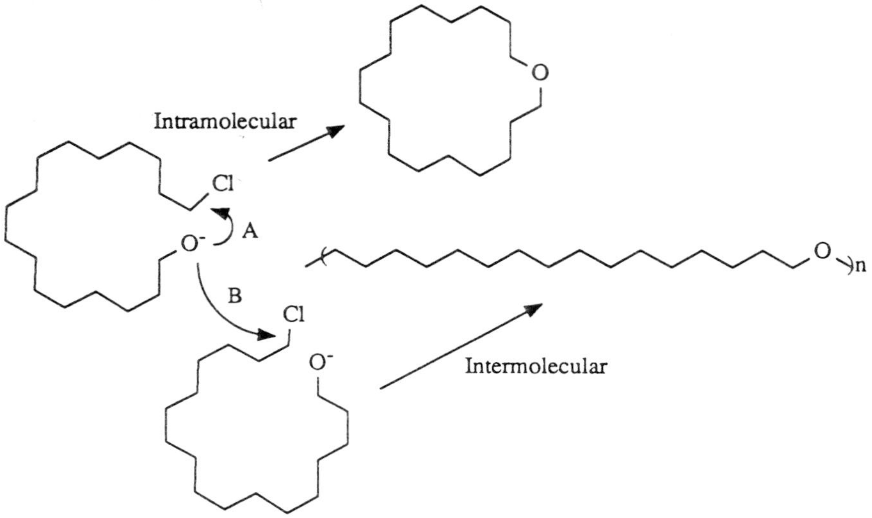

Figure 2.4 *The principle of the template effect*

cation present in the reaction mixture. Since there is a polar interaction between the negative oxygens and the positive cation, they tend to organize themselves around the cation. By forming a semicircle or circle about the cation, the likelihood of the two ends being near each other is greatly increased.

In Scheme 2.3, note that one intermolecular bond forms in the absence of any template effect. This leads to a hexa(ethyleneoxy) system having an alkoxide residue at one end and a tosylate leaving group at the other. The sodium cation has an affinity for the oxygen atoms and organizes the intermediate about it. Thus the pole–dipole interactions between Na^+ and oxygen (and pole–pole interaction with the terminal O^-) bring the ends of the reacting molecule into the proximity required for ring closure. Thus, the template effect favors the first order pathway over the second order process.

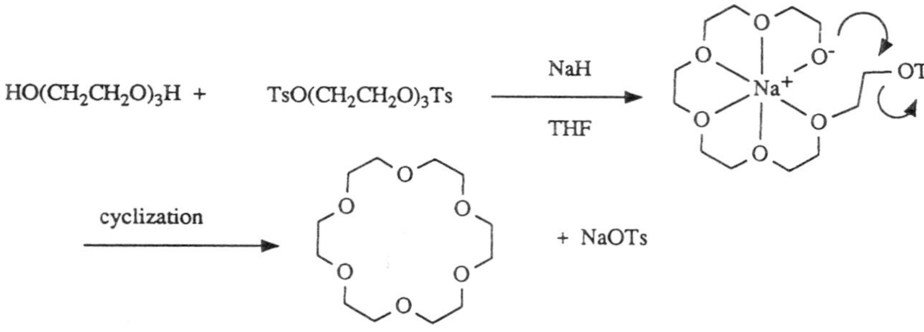

Scheme 2.3

There can be little doubt that the template effect is a real phenomenon but it must also be admitted that the process has not been investigated systematically in many cases. Because there is good and convincing evidence that the template effect operates in some cases, it has often been assumed that it operates in all cases. On the other hand, suggestions have been made that the effect has more to do with base strength than complexation. Moreover, in certain instances where no templating cation was present, high yields were obtained anyway. As sometimes happens in chemistry, many workers in the macrocycle field have moved to new horizons without the formulation of definitive rules concerning this important principle.

The first real evidence for the template effect came from Greene's work. Pedersen had previously synthesized 18-crown-6 by using hexaethylene glycol monochloride and potassium *tert*-butoxide. The yield in the cyclization was only 1.8%. Greene developed a synthesis using triethylene glycol and triethylene ditosylate and showed that the yield was dramatically increased by the presence of metallic cations, particularly potassium. It was already known that 18-crown-6 is selective for potassium cation and by these studies Greene demonstrated the applicability of the template idea to macrocyclic polyether synthesis. As noted above, Busch had already demonstrated the importance of the template in the synthesis of nitrogen macrocycles. Chastrette and Chastrette had shown that the reaction of furan with acetone, a process known for some years (see Section 1.2.3), gave an increased yield of the tetrameric macrocycle when lithium cation was present (Scheme 2.4).

Liotta and his co-workers examined cyclization reactions involving such small ring systems as 12-crown-4 and showed that the presence of lithium cation improved the yield. Okahara and co-workers reported a temperature dependence study of cyclization and Reinhoudt and co-workers also contributed in important ways to our knowledge of the template effect. In recent years, the most important contributions in this area have probably been from the Italian school of Mandolini and associates.

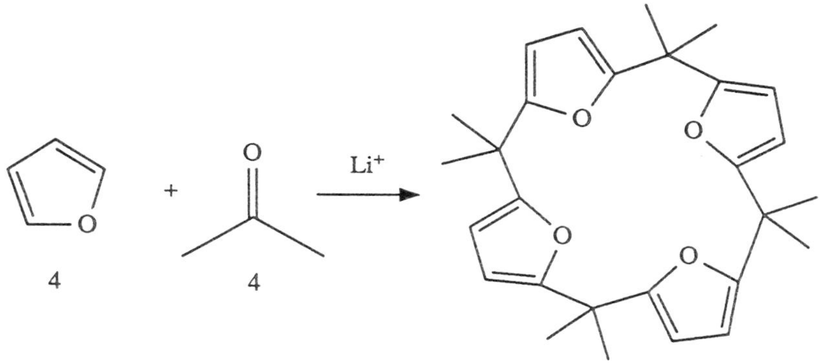

4 + 4 $\xrightarrow{Li^+}$

Scheme 2.4

Using an intramolecular phenolic nucleophile and a variety of cationic templates, Mandolini and Masci demonstrated in a convincing fashion that the template effect operates in their case (Scheme 2.5).

There is one other set of experiments that was influential in connection with the template effect. The cyclo-oligomerization of ethylene oxide using aluminum sesqui-halides was patented in England in the 1950s (see Section 1.2.2). Dale (pronounced Dah'-leh) showed that BF_3 could be used as the catalyst (Scheme 2.6). In this case, added cations changed the ratios of cyclic products depending upon cation size. For example, when sodium cation was added to the reaction mixture, a greater proportion of 15-crown-5 (**12**) was obtained than when a larger cation was present.

This work was reported in communication form with few experimental details. The patent that was eventually issued on this process contained greater detail and showed that most of the cyclic product is dioxane (**13**) whether or not a cation is present. Nevertheless, it is clear that the cation does alter the composition of cyclic product. This was interpreted in terms of a template or organizational effect. A simple template effect still seems to be the most likely explanation for this phenomenon. Unfortunately, because dioxane is the preponderant product of this process, it has not proved to be an attractive synthetic procedure, at least for most academic laboratories.

There are still cases where the expectations of the template effect are unfulfilled or not understood. These are beyond the scope of this monograph. Suffice it to say, however, that the evidence for a template effect is, in many cases, irrefutable. Invocation of a template effect should, however, be done with some caution.

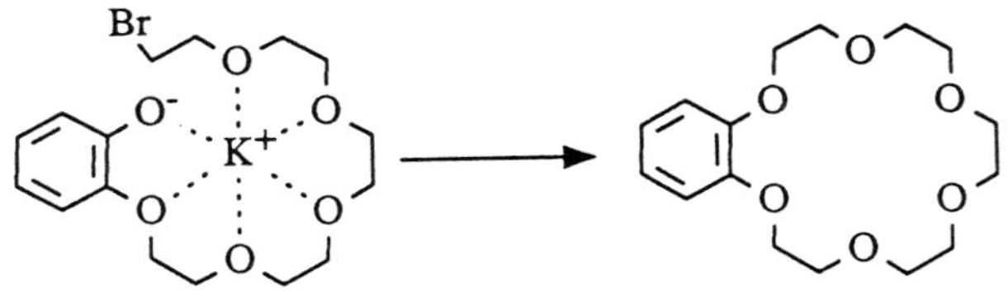

Scheme 2.5

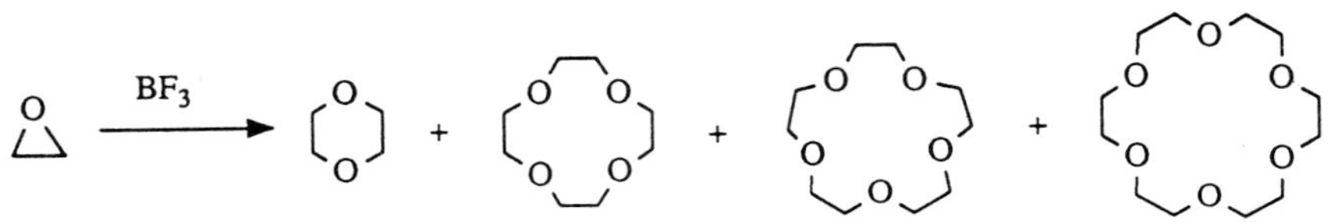

13 **12**

Scheme 2.6

2.5 Early Syntheses of Crowns

2.5.1 Dicyclohexano-18-crown-6

The synthetic approach Pedersen used to obtain dibenzo-18-crown-6 has already been recounted (see Section 2.2). Pedersen also showed that hydrogenation of dibenzo-18-crown-6 (high pressure and a ruthenium catalyst) afforded dicyclohexano-18-crown-6 as a mixture of isomers (Scheme 2.7). The latter compound was more lipophilic and a better complexing agent because all of its oxygen atoms are *sp*[3] hybridized. To this day, dicyclohexano-18-crown-6 remains an important candidate for use when binding strength *and* lipophilicity are required. The only drawback is that all possible stereoisomers form.

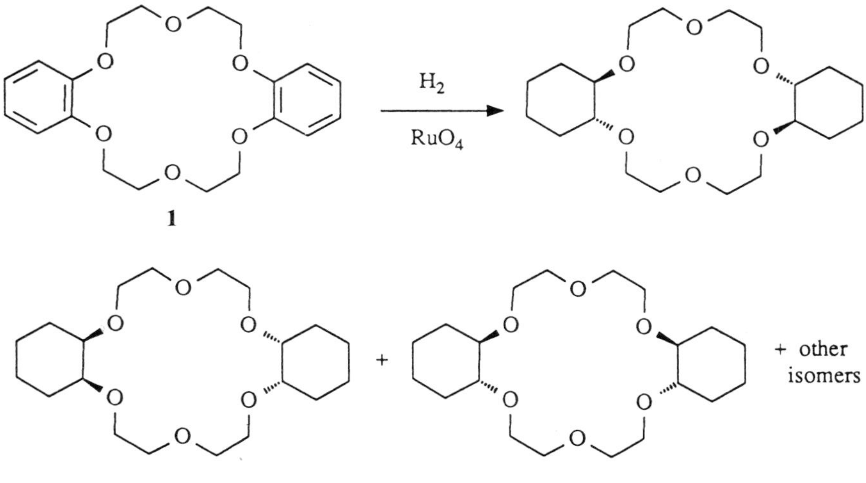

1

H_2

RuO_4

\+ other isomers

Scheme 2.7

2.5.2 18-Crown-6

In addition to his cyclo-oligomerization studies, Dale showed that 18-crown-6 could be prepared from triethylene glycol ditosylate and triethylene glycol in benzene solution. Dale obtained pure 18-crown-6 after vacuum distillation. It now seems remarkable that in those days it was unclear whether crowns could survive distillation. Distillation was a key step in making 18-crown-6 readily available. Pedersen, Greene, and Dale had already shown that 18-crown-6 could be prepared under a variety of conditions. The Author, working in Cram's laboratory in the early 1970s, felt it might be possible to prepare these crowns simply by stirring together triethylene glycol dichloride, triethylene glycol, and potassium hydroxide in THF solution (Scheme 2.8). Water, 10% by volume of the

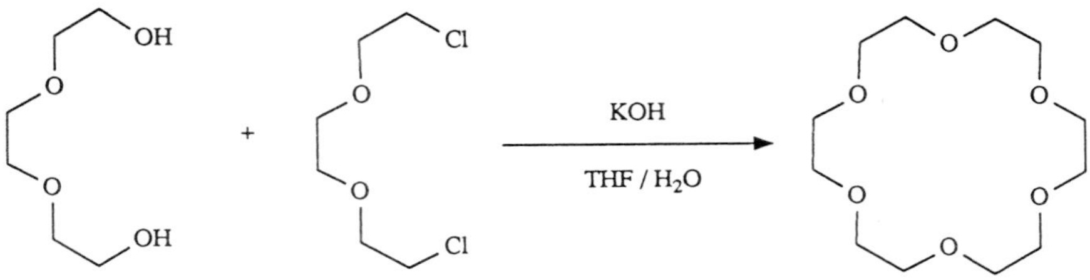

KOH

THF / H_2O

Scheme 2.8

mixture, was added to help the KOH dissolve. This approach offered both experimental simplicity and economy.

By using this procedure, material of modest purity could be obtained after a single vacuum distillation. Liotta and co-workers had obtained a copy of the unpublished and preliminary procedure. They observed that when the crude reaction mixture was diluted with acetonitrile, a precipitate formed. This solvate proved to be the key to obtaining 18-crown-6 in very pure form. The procedure is now an *Organic Syntheses* preparation and 18-crown-6 can be obtained reproducibly in pure form in 25% overall yield.

2.5.3 [2.2.2]-Cryptand

From the structural point of view, the major difference between crowns and cryptands is the presence of a third chain that converts normally two-dimensional crown ethers into a three-dimensional ligand system. 18-Crown-6 (**14**) and [2.2.2]-cryptand (**15**) are representative of each group.

[2.2.2]-Cryptand is based upon an 18-membered ring just as is 18-crown-6. Whereas Pedersen used nucleophilic substitution reactions to prepare most of the crown ethers he reported, Lehn and co-workers used high dilution cyclization reactions involving amide formation. Indeed, two successive macrocyclizations were involved: the first formed the 18-membered ring and the second added the third chain (Figure 2.5).

In Lehn's approach to these molecules, triethylene glycol was the precursor to triglycolic acid, $HOOCCH_2OCH_2CH_2OCH_2COOH$ which, in

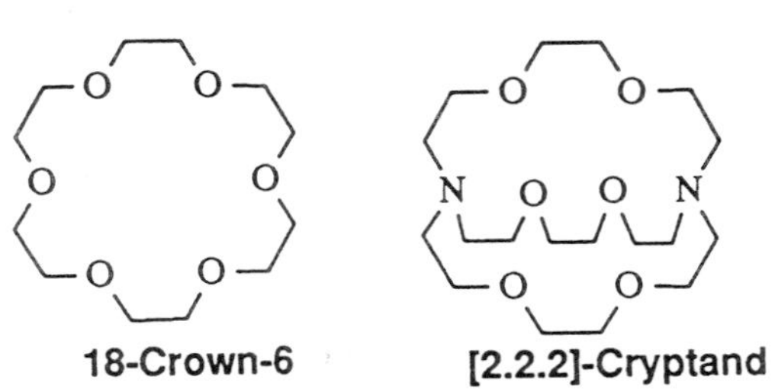

18-Crown-6

14

[2.2.2]-Cryptand

15

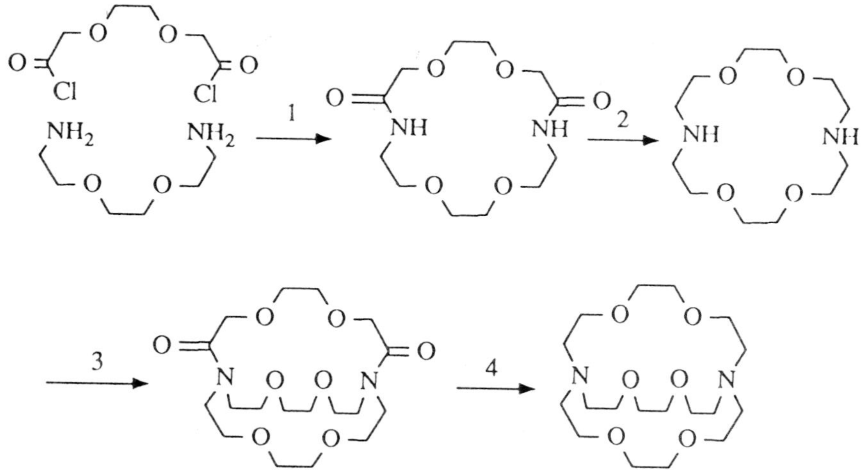

Figure 2.5 *Synthetic scheme for [2.2.2]-cryptand. Step 1: cyclization at high dilution. Step 2: hydride reduction. Step 3: cyclization as in step 1. Step 4: hydride reduction*

turn, afforded the dichloride. Cyclization was conducted using high dilution conditions in both steps 1 and 3. Because the formation of diglycolic acid is somewhat difficult, and because four steps are required for the synthesis, cryptands remain much more expensive than crowns ethers, even today.

2.6 Syntheses of Crown Ethers

The three basic decisions that must be made about any crown ether system are ring size, the presence of heteroatoms other than oxygen, and whether or not ancillary functions such as sidearms, sub-cyclic units, *etc.* are required. Syntheses of macrocycles are discussed below depending on these choices, especially the presence of heteroatoms.

2.6.1 Monomers, Dimers, and Trimers

The synthesis of dibenzo-18-crown-6 has been recounted in Section 2.2. One aspect of the preparation requires some elaboration, however.

The synthetic approach to dibenzo-18-crown-6 is sometimes called a 'shotgun' reaction because catechol and 2,2′-dichlorodiethyl ether each have two reactive sites. Thus, they may react on a 1:1 basis to form benzo-9-crown-3 (monomer), 2:2 to afford dibenzo-18-crown-6 (dimer), and so on. The templating effect of Na^+ appears to favor the 18-membered ring. Larger cations favor a greater proportion of larger ring(s). It is important to recognize that doubling the 1:1 ratio of reactants in the hope of favoring a 2:2 (dimeric) product is futile as 2:2 is equivalent to 1:1. Changing the cation, the concentration, or other variables may have the desired effect but using equivalent stoichiometries will invariably lead to equivalent product mixtures. The monomer (**16**), dimer (**1**), and trimer (**17**) products arising from the reaction of catechol with diethylene glycol dihalide are shown.

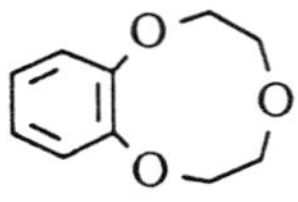

16

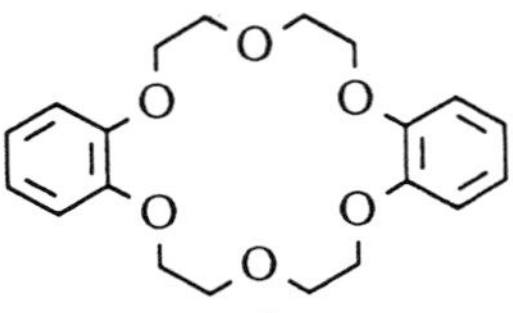

1

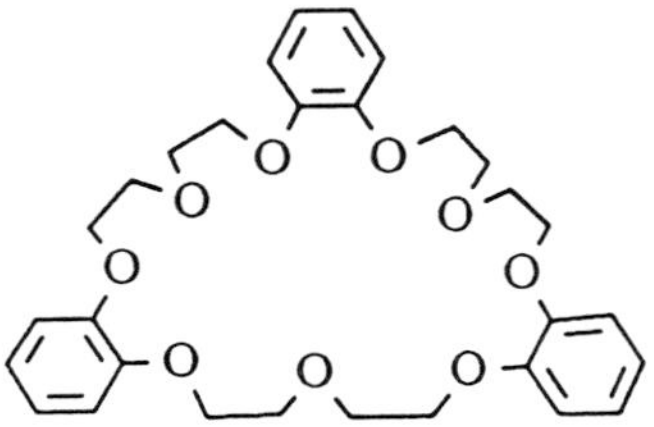

17

It is amusing to note that more than one inexperienced but otherwise fine co-worker in the Author's laboratory either attempted this trick and failed or could not understand why the yield of a 2:2 reaction was never greater than 50%.

2.6.2 Nucleophiles and Bases

If there is a basic reaction in crown ether synthesis, it is the Williamson reaction. The Williamson ether synthesis involves an oxygen nucleophile that displaces a leaving group either on another molecule or at another position within the same molecule. An ether is formed in the process. If hexaethylene glycol monochloride (**18**) cyclizes by this reaction, then the oxygen nucleophile at one end of the molecule displaces chloride at the other end affording 18-crown-6 (**14**) as product.

If triethylene glycol is the nucleophile and triethylene glycol dichloride is the electrophile, then a single nucleophilic substitution reaction will afford the starting material for the reaction shown above: hexaethylene glycol monochloride. A second nucleophilic substitution will then afford 18-crown-6. Both processes involve the Williamson reaction although the latter sequence employs two such steps (see Scheme 2.8).

In the general sense, a crown ether is a poly(oxygen) macrocycle and we will thus restrict the discussion to the question of oxygen nucleophiles. The key question becomes whether the oxygen nucleophile is an alkoxide or a phenoxide. The alkoxides are somewhat more difficult to generate and slightly less nucleophilic than are the phenoxides. Overall, though, the difference is small and both have successfully been employed as nucleophiles in a broad range of macrocycle syntheses. Reactions involv-

$$\underset{\mathbf{18}}{\text{[structure]}} \xrightarrow{K^{+-}O\text{-}t\text{-}Bu} \underset{\mathbf{14}}{\text{[structure]}}$$

ing phenolic nucleophiles often utilize relatively weak bases such as sodium or potassium hydroxide. Historically, the choice of base (*i.e.* choice of cation) was dictated by the size of ring desired. It is now known that potassium cation binds most simple crowns more strongly than does sodium. Potassium ion would probably be a better template ion for many known crown syntheses although, to the Author's knowledge, no systematic study verifies this.

The hydroxide bases are sufficiently powerful to deprotonate phenols (and even some aliphatic alcohols). Stronger bases can be used to generate phenoxide anions but this is usually not necessary. The use of a stronger base may not be detrimental, except possibly in the economic sense. The deprotonation of phenol by hydroxide yields phenoxide and the by-product is water. Water may be a problem in crown synthesis since hydroxide itself may displace the leaving group.

Deprotonation of alkoxides by hydroxide is less efficient than the corresponding reaction of phenols. The pK_a range for saturated alcohols is 16–19 ($K_a = 10^{-16}$–10^{-19}) and the pK_a of water of 16. The best one can hope for is an equilibrium between hydroxide and alkoxide. For this reason, the much more powerful base sodium hydride has been a common choice for the preparation of aliphatic crown ethers. It is likely that potassium hydride, because of potassium's binding behavior, would be a more effective base but it is more difficult to handle than sodium hydride. The practical issues of availability and handling thus overwhelm the potential value of potassium as a template ion. A wide variety of other systems such as dimsyl sodium, sodium or potassium *tert*-butoxide, *etc.* have been applied in crown ether preparations. Even so, sodium hydride remains the most used base for aliphatic crown syntheses.

2.6.3 Leaving Groups

The question of leaving groups in the Williamson reaction is likewise a compromise of economy, efficiency, and the ability to obtain the substrate in the purest form. The principal choices are halides, mesylate, and tosylate. Among these, the lowest molecular weight and least expensive leaving group is chloride. When it can be used, it is the obvious choice. Bromide and iodide are more reactive alternatives but the increase in molecular weight is significant as is the increased tendency to oxidize. Iodine has also been applied extensively as a leaving group but this has mostly been in azacrown syntheses. In these cases, the iodide leaving group is often used in conjunction with sodium iodide, a combination that works although the influence of added NaI in the presence of the iodide leaving group is not understood.

Alkyl chlorides are often distillable whereas the tosylates rarely are. On the other hand, tosylates are often crystalline whereas the halides rarely are. Iodide gives greater reactivity than chloride but the iodides are more prone to decomposition than are the chlorides. Further, the iodides have much higher boiling points than do the chlorides and purification of these materials is correspondingly more difficult (see Figure 2.6).

Reactivity: $TsO^- \geq I^- > Br^- > Cl^-$

Boiling points: $I^- > Br^- > Cl^-$

Tosylate: does not distill but usually crystalline

Figure 2.6 *Comparison of leaving groups*

The consideration of economy aside, the most popular leaving group has proved to be tosylate. The advantage of the tosylate group is really twofold. First, it is usually very easy to synthesize a tosylate from a primary alcohol. [The primary alcohol dissolved in pyridine is added to a solution or slurry of tosyl chloride in the same solvent at 0 °C. After standing for about an hour, the product is obtained by filtration, extraction, and crystallization (Scheme 2.9).]

Second, the primary tosylates often are solids and can be recrystallized to a high level of purity. This advantage is especially important with the low molecular weight oligoethylene glycols. Diethylene glycol ditosylate (available from diethylene glycol), for example, is a beautifully crystalline material melting at 78 °C. It can be prepared in about two hours and recrystallized to purity from 95% ethanol.

A critical but not widely recognized problem is that oligoethylene glycols often contain homologs as contaminants, *i.e.* $HO(CH_2CH_2O)_nH$ contains some $HO(CH_2CH_2O)_{n+1}H$ or $HO(CH_2CH_2O)_{n-1}H$ (Scheme 2.10). If the crown ether is prepared from a contaminated mixture, and the ditosylate cannot be crystallized to purity, it is very difficult to remove the homologous crown(s) after the cyclization reaction has been completed. An example of this problem is found with tetraethylene glycol, a compound for which the ditosylate is not crystalline. Commercial samples of tetraethylene glycol sometimes contain 10% or more of higher and lower

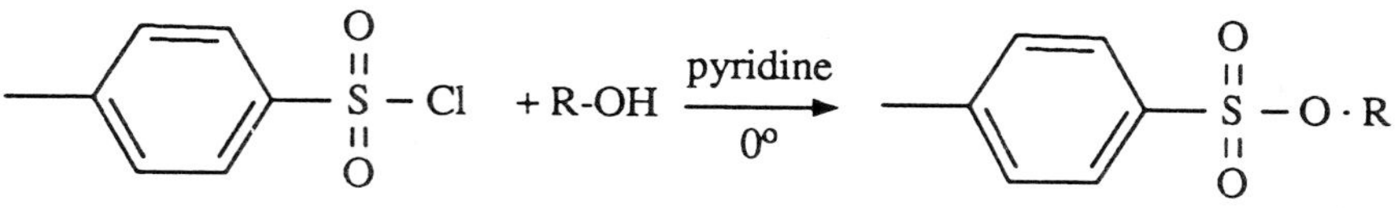

Scheme 2.9

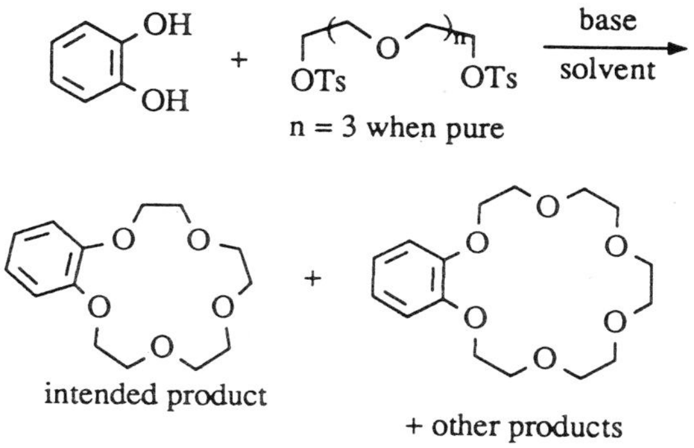

Scheme 2.10

homologs. Fortunately, however, tetraethylene glycol can be distilled (spinning band or Vigreux column) to afford more than 99% of the tetramer, $HO(CH_2CH_2O)_4H$. In this case, use of the tosylate does not solve the homolog problem but its use is still advantageous because the diol can be so readily converted into it. The preparation of tetraethylene glycol dichloride is economically preferable for large scale applications but the preparation is more cumbersome, at least in the academic laboratory. Most diols that afford ditosylates can also be converted into dichlorides so the successful synthesis can often be economized when scaled-up.

Mesylate, CH_3–SO_2–, is a useful alternative to tosylate. Mesylate offers some of the advantages of the tosylate and is roughly its equal in nucleofugacity. Mesylate derivatives are usually less crystalline than tosylates but they are lower molecular weight than tosylate. The latter means that a smaller quantity of material must be handled. On small scales, this is a marginal advantage. Mesylate usually offers no particular advantage in terms of yield.

In sum, tosylate is probably the preferred leaving group for exploratory reactions and has probably been used in the largest number of cases. In terms of volume, however, chloride is probably preponderant since it is this that is preferred for economic reasons.

2.6.4 Solvents

We noted briefly above that when a dianion such as catechol is used, a reasonably polar solvent is required in order to solubilize the dianion. The choice of nucleophile (anion) dictates the choice of solvent. For example, to use sodium hydride as a base and ethanol as solvent would be useless. The reaction between sodium hydride and ethanol would afford sodium ethoxide instead of, or in equilibrium with, the desired nucleophile (anion).

Often, specific bases command specific solvents. For example, potassium *tert*-butoxide is usually formed in and used in *tert*-butyl alcohol. Potassium *tert*-butoxide is now available commercially and could therefore be used in other solvents. Since $K^{+-}O$–*t*-Bu is more soluble in tetrahydrofuran (THF) than in *tert*-butanol, it is often used in the ether rather than the alcohol solvent. The use of this base in crown synthesis still poses the problem of *tert*-butoxide competing with the desired nucleophile. The latter problem is greater when *tert*-butoxide is used in THF rather than in *tert*-butanol.

Sodium hydride (NaH) reacts readily with dimethylsulfoxide (CH_3–SO–CH_3) to produce dimsyl sodium (CH_3–SO–$CH_2^-Na^+$). Dimsyl sodium is a powerful base and can be used in solvents other than DMSO. The combination of NaH/DMSO has found little application in crown syntheses, probably because the solvent is difficult to remove from the reaction mixture. The combination of NaH and DMF has proved to be a more common choice when a polar solvent is required.

If sodium hydride is the preferred base for macrocycle syntheses using the Williamson reaction, then THF and DMF are the preferred solvents. THF is less polar than DMF but it is much easier to remove from the reaction mixture. Furthermore, THF does not hydrolyze as DMF is inclined to do under vigorous, basic conditions. DMF is difficult to remove even when a reasonably high vacuum is used for evaporation. It always seems that there is some solvent residue that must be removed by extraction. Final purification of a DMF reaction often involves an acid wash as well. All these features are disadvantages but DMF remains popular because its polarity is critical to the success of many crown syntheses.

The Williamson reaction is fundamentally different from high dilution cyclization. Benzene has been a particularly favored solvent for high dilution reactions, especially in academic laboratories. Safety restrictions preclude the use of benzene in many industrial laboratories but toluene may be a useful substitute for it. Although some Williamson reactions have been conducted in benzene, it has not been a favored solvent because most alkoxide anions are poorly soluble in it. Interestingly, benzene may actually be advantageous in this respect. Notwithstanding the template effect, high dilution will invariably favor intramolecular *versus* intermolecular reaction. If the solubility of the alkoxide nucleophile is minimized, then its concentration is likewise minimized and the effective dilution is increased.

The issue of what solvent should be used in any given bimolecular process is interesting. For a second order process, the rate depends on both nucleophile and electrophile concentration. The highest possible concentration of each reagent can be achieved if no solvent is added. On the other hand, high polarity solvents can increase the rate of a polar reaction. When solvent is added, there must be some loss of reactivity due to reduced concentration, but this may be more than compensated by the polarity-induced rate increase.

It might therefore be advantageous to use a solvent such as hexamethylphosphoramide {HMPA or HMPT, $[(CH_3)_2N]_3PO$} for all S_N2 reactions. In practice, this solvent is little used since it is toxic and its boiling point is so high that it is difficult to remove from the reaction mixture. Other dipolar aprotic solvents, however, afford advantages without the same problems. Some common solvents that should be considered for crown synthesis are tabulated (Table 2.1).

2.6.5 High Dilution

The obvious way to conduct a reaction at high dilution is to use a small amount of each reactant and a huge amount of solvent. This is not usually economically feasible and it is often impractical. Methods have been developed so that high dilution processes may be conducted in a reasonable fashion. The most common method currently in vogue is to use a relatively small reaction volume and slowly add the reactants simultaneously from two addition funnels. If the reaction rate is greater

Table 2.1 *Common dipolar aprotic solvents*

Structure	*Name and Abbreviation*	*Boiling Point °C*	*Water solubility*
CH_3COCH_3 [a]	Acetone	56	miscible
$(CH_2)_4O$	Tetrahydrofuran, THF	65	miscible
$CH_3C{\equiv}N$	Acetonitrile, MeCN	81	miscible
CH_3NO_2 [b]	Nitromethane	101	9.1/100g
$(CH_3)_2N\text{-}CHO$	Dimethylformamide, DMF	153	miscible
$CH_3\text{-}SO\text{-}CH_3$	Dimethylsulfoxide, DMSO	189	miscible
$H\text{-}CO\text{-}NH_2$	Formamide	210	miscible
$(CH_3)_2N\text{-}PO[N(CH_3)_2]_2$	Hexamethylphosphoramide HMPT, HMPA	230	miscible

[a] The pK_a of acetone is approximately 20 and it may be deprotonated by the bases used in crown syntheses. [b] The pK_a of nitromethane is approximately 10 and it is readily deprotonated by NaOH

than the addition rate, the reactant concentration is kept low and the dilution is therefore high.

When the Author was a graduate student, the compound cyclam (tetraaza-14-crown-4) was required by an inorganic colleague. The Author was asked to synthesize this molecule according to a published procedure. The reaction was conducted in a nearly full twelve liter (12 000 ml) flask and only 1–2 g of product were eventually obtained. Cyclam can now be synthesized using a template effect and also using slow addition of reagents. Templated cyclizations can now be conducted in far smaller glassware and the products obtained in much higher yield. This is important because handling 5–10 l of solvent in order to obtain 1–2 g of material is burdensome and ultimately impractical, not to mention physically demanding. Moreover, if procedures for the syntheses of compounds are impractical, the compounds are less available and usually less well studied.

The approach described above may be automated by use of a syringe pump. These devices are more or less ordinary syringes, the plungers of which are motor driven. The speed may be adjusted so that two syringes release the same volume of reagent over a long period of time. It is quite feasible to add, for example, 20 ml of each reagent solution during a period of 8 hours (4.2×10^{-5} l min^{-1}) with precision using such a device.

2.6.6 General Conditions

We have noted that many variations in synthetic procedure are available for macrocycle syntheses. It is probably unwise to offer general rules as these can always be contradicted by at least one example. In the Author's laboratory, a new crown synthesis is likely to be performed as follows, at least in the first attempt.

Sodium hydride (NaH) as base
Anhydrous THF as solvent
The diol as nucleophile
The ditosylate or dimesylate as electrophile
Vigorous mechanical stirring, reflux temperature
Purification by filtration, chromatography, and finally bulb-to-bulb distillation or crystallization

2.7 Incorporation of Aromatic Subunits

One of the most important substitutions that has been made in simple macrocycles is the incorporation of aromatic subunits in place of one or more ethyleneoxy units. The most obvious substitution is catechol (1,2-dihydroxybenzene) for a single ethyleneoxy unit. Numerous other substitutions have been carried out as well. These substitutions are numerous indeed, but generally uncomplicated. Rather than discussing each case, we show a few examples in the hope that they adequately represent the imagination that led to their syntheses (structures **19–26**).

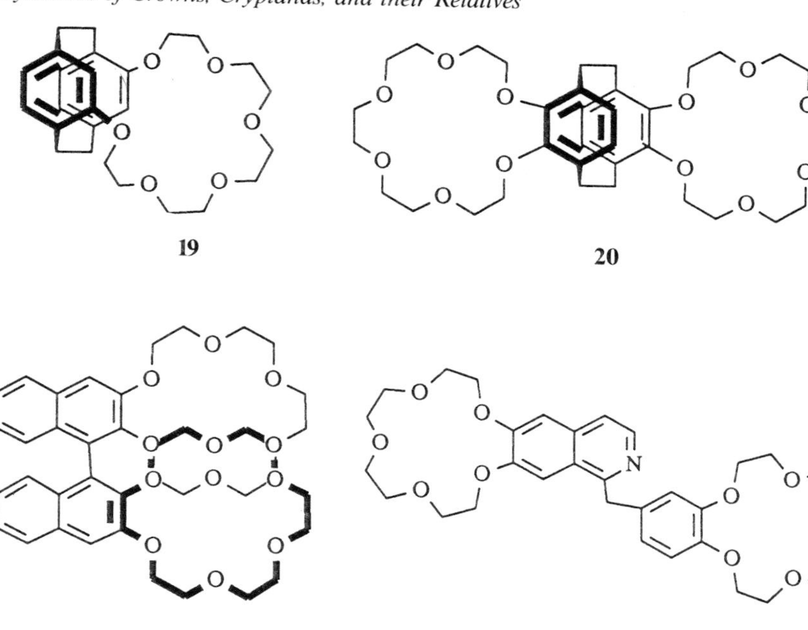

19 **20**

21 **22**

23 **24**

25 **26**

Perhaps one of the best known groups of structures involves incorporation of the 1,1′-binaphthyl group by Cram and co-workers. This unit confers chirality on the macroring system. Two isomers of dibinaphtho-22-crown-6 are shown in structures **27** and **28**. The ring size designation '22' is obtained by counting *all* interior atoms in the largest ring. The use of such structures in resolutions is discussed in Chapters 3 and 5.

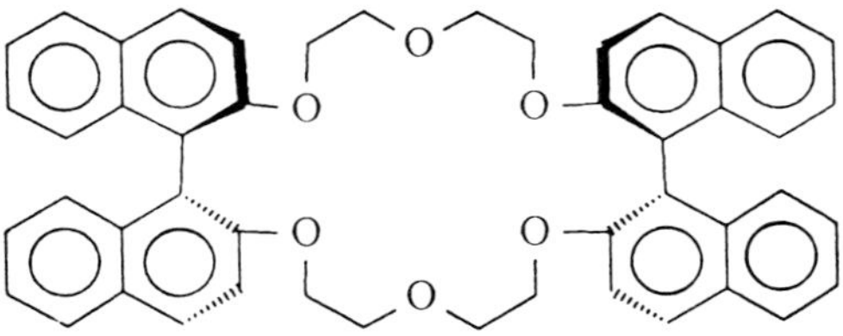

27

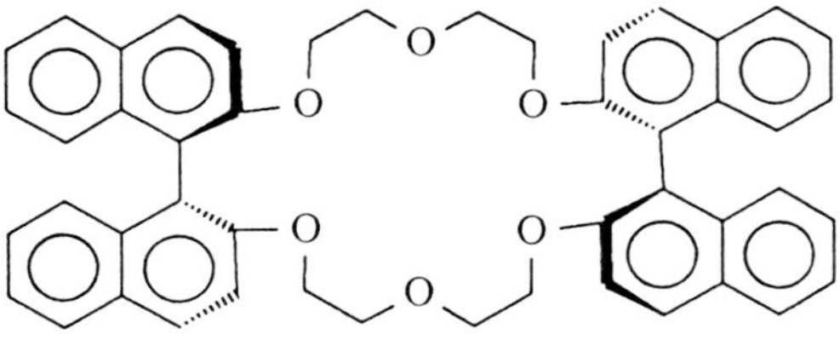

28

2.8 Nitrogen-containing Macrocycles

To the organic chemist, there is no more obvious substitution than nitrogen for oxygen in a heterocyclic structure. Indeed, Pedersen substituted both nitrogen and sulfur for oxygen in a number of the earliest compounds he prepared. In substituting nitrogen for oxygen, the size and shape of the macrocycle remain similar (although the conformations may change) but the acid–base properties of the material are substantially altered. Moreover, the cation binding selectivity is altered as well because nitrogen and oxygen have inherently different affinities for cations.

The preparation of macrocycles containing even one nitrogen introduces a set of problems different from those encountered for the all-oxygen macrocycles. In the latter case, the key difference was whether the oxygen nucleophile was phenoxide or alkoxide. In the case of nitrogen-containing macrocycles, we must consider if nitrogen will be tertiary. If not, it must be protected in some stage of the synthesis. The various problems and some current methods for resolving these difficulties are discussed below.

2.8.1 Protection of Nitrogen

Some of the earliest azacrown syntheses were attempted by Lockhart and her co-workers. They found, for example, that the reaction of *ortho*-aminophenol (**29**) with tetraethylene glycol dichloride (**30**) afforded a product mixture (Scheme 2.11). Although both compounds are interesting, methods for protecting nitrogen were desireable so that the 15-membered ring structure would be formed preferentially.

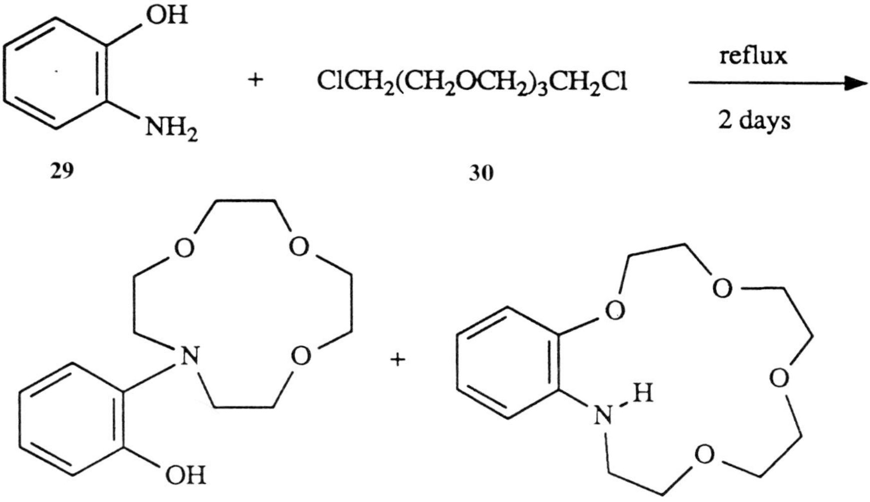

$ClCH_2(CH_2OCH_2)_3CH_2Cl$ (**30**), reflux, 2 days

Scheme 2.11

An obvious means for protecting nitrogen utilizes the chemistry inherent in the well-known 'Hinsburg test.' An amine, R–NH_2, has a pK_a of about 33–35 for the N–H group and is very weakly acidic. When treated with an arenesulfonyl chloride, Ar–SO_2–Cl, the N–H group on the sulfonamide, Ar–SO_2–NH–R, now exhibits a pK_a of about 17, an increase in acidity of nearly 20 powers of ten. The sulfonated amine is therefore much easier to deprotonate and readily forms an anion (nucleophile). The other and greater advantage is that only one of the original two protons remains so that only one new N–X bond will form. The arenesulfonamide group therefore both protects one of the nitrogen N–H protons and acidifies the other.

The disadvantage of the sulfonamide protecting group is that it is often difficult to remove after the reaction is over. Numerous conditions have been developed for its removal and no doubt they work as reported. None of the methods is either particularly mild or very general and often no reaction is observed or removal of the protecting group is accompanied by much loss of material.

The synthesis of aza-18-crown-6 affords an example of this difficult problem and one early solution to the protection problem. When this molecule was first required in the Author's laboratory, during the mid-1970s, the obvious approach at the time seemed to be to protect nitrogen as its tosylamide. Diethanolamine (**31**) could be mono-*N*-tosylated as in **32** or tritosylated as in **33**. The latter approach was tried but it was found that elimination was a consistent problem. Since nitrogen is more nucleophilic than oxygen, clean monotosylation was possible using Na_2CO_3 as base.

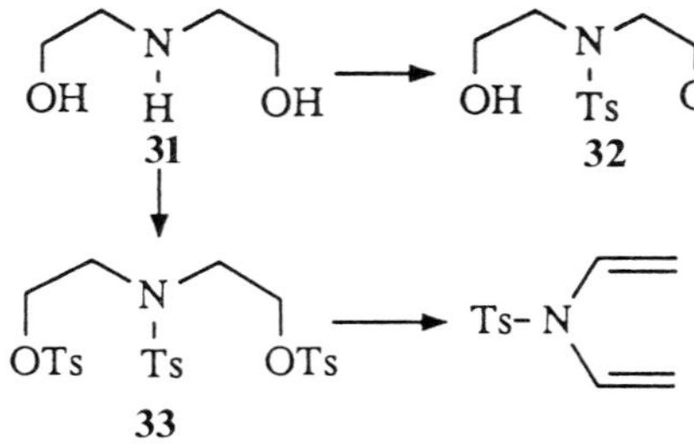

In the Author's early efforts to synthesize aza-18-crown-6 (**34**), the tosylamide was favored as a protecting group for diethanolamine. It made the starting material quite crystalline and thus it was readily purified. The cyclization proceeded well and in good yield. Unfortunately, the *N*-tosyl-crown residue resisted all efforts to remove the protecting group. Normal hydrolytic conditions involving both acid and base were attempted as were use of strongly nucleophilic systems such as isoamyloxide in tetrahydrofuran under pressure at 180 °C.

To be sure, removal of the tosylamide group has been effected in some cases. These successes have usually involved reductive methods or the use of strongly acidic solutions. Early work on polynitrogen macrocycles by Atkins and Richman involved removal of the tosyl groups with concentrated sulfuric acid (Scheme 2.12). Although these conditions may seem quite drastic, they were effective for the cyclams. Obviously, in compounds containing sensitive functional groups, this approach would be less appropriate.

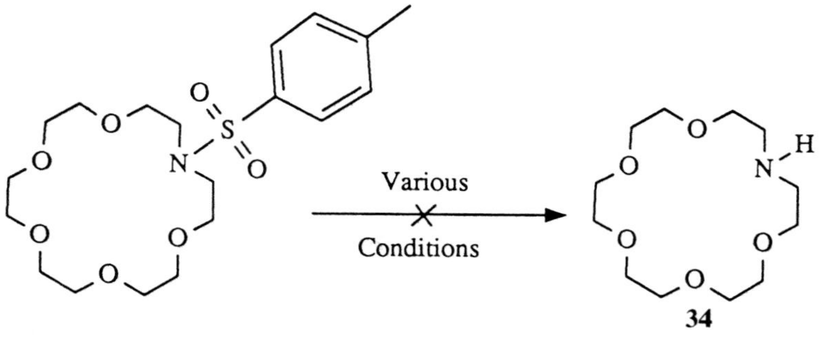

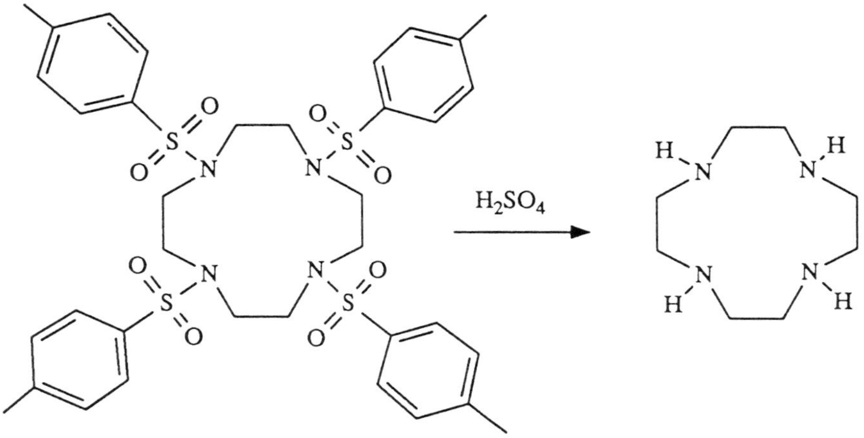

Scheme 2.12

Because arenesulfonyl protecting groups are often hard to remove, other protecting groups have been favored. The dominant protecting group for nitrogen has become benzyl. The benzyl group may be used to alkylate nitrogen in the presence of hydroxyl oxygen with high selectivity and it can be removed readily by hydrogenolysis. The benzyl group does not enhance crystallinity but it also does not increase molecular weight too much. Thus, the prospect of purification by crystallization is not increased, but distillation is not hampered. Removal is usually accomplished by hydrogenation of an ethanol solution of the macrocycle at 3–4 atm. in a Parr 'shaker' device. Occasionally, removal of the benzyl group has proved unexpectedly difficult. In such cases, the solution is usually to add a trace of concentrated HCl. The preparation of aza-18-crown-6 (**34**) can thus be accomplished as in Scheme 2.13.

Other groups, such as benzhydryl and trityl, have been used occasionally to protect nitrogen. The advantage of these groups over benzyl is in ease of removal. The difficulty, of course, is that they are higher molecular weight and thus add bulk to the reagent. They are generally more expensive as well.

Another azacrown synthesis that deserves mention was developed by Okahara and co-workers. This cleverly incorporates several features of other syntheses. A primary amine, $R{-}NH_2$, reacts with ethylene oxide to afford a diethanolamine derivative, $R{-}N(CH_2CH_2OH)_2$. This is, in turn, treated with more ethylene oxide to form the precursor **35**. This is then treated with tosyl chloride which tosylates one hydroxyl *in situ* leading to cyclization (Scheme 2.14).

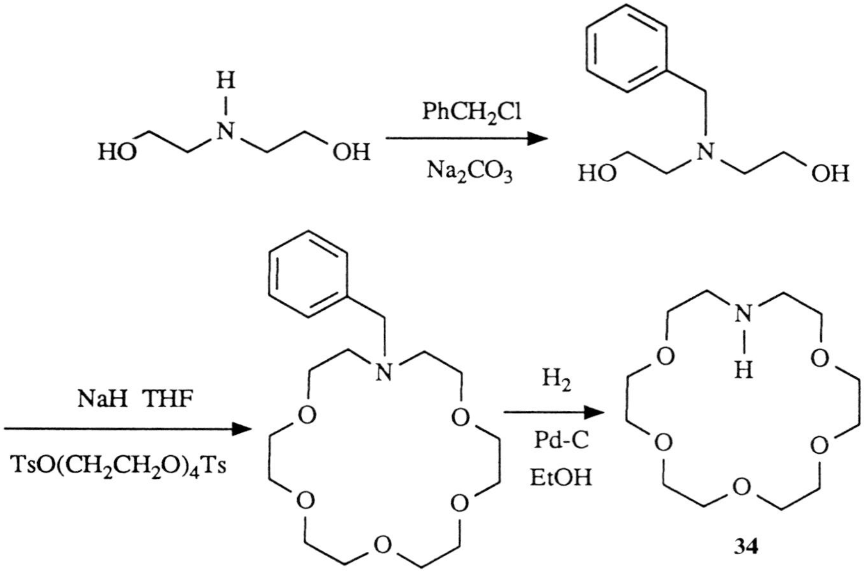

$PhCH_2Cl$

Na_2CO_3

NaH THF

$TsO(CH_2CH_2O)_4Ts$

H_2

Pd-C

EtOH

34

Scheme 2.13

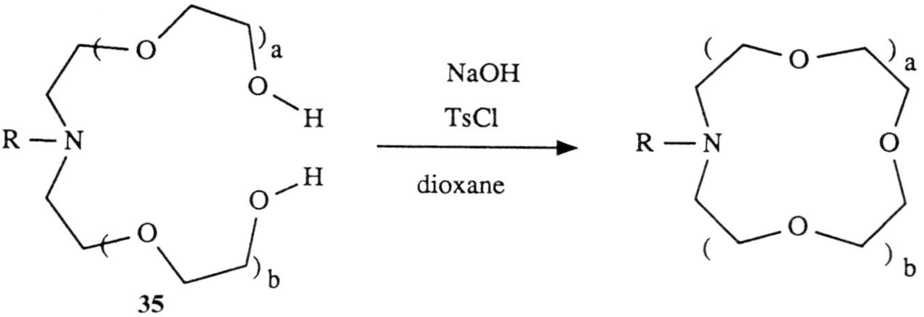

NaOH

TsCl

dioxane

35

Scheme 2.14

2.8.2 Lactam Formation Followed by Reduction

A second important strategy for azacrown compound synthesis is to form a cyclic amide (lactam) or *bis*(lactam) and then reduce to the corresponding amine. In this approach, an acid chloride is treated with a primary amine under high dilution conditions. No protecting group *per se* is required. As soon as the amide link is formed, the nitrogen atom is protected by resonance. A further acylation (or alkylation, for that matter) is not possible because of the partial positive charge on nitrogen (see Scheme 2.15).

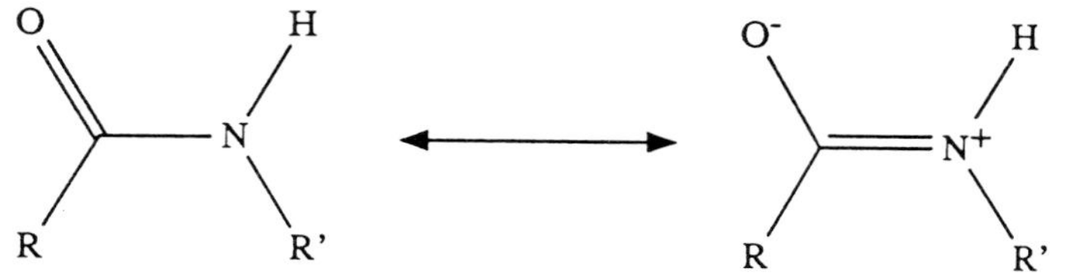

Scheme 2.15

Indeed, this mode of synthesis has been widely used for the preparation of dinitrogen heteromacrocycles. Such an approach was used by Lehn and co-workers in the early synthesis of diaza-18-crown-6 (**36**), a key intermediate required for cryptand preparation. This reaction requires a diacid chloride and a diamine.

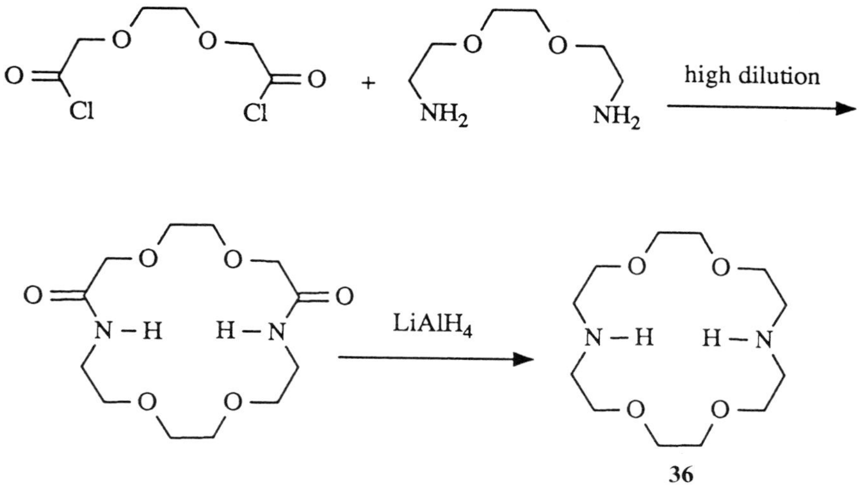

high dilution

$LiAlH_4$

36

Both diborane (B_2H_6 or 'BH_3') and lithium aluminum hydride (LAH, $LiAlH_4$) reduce amides to amines in a variety of structural circumstances. In some groups there is a clear preference for diborane as a reducing agent and it has generally proved to be the better choice in experiments conducted in the Author's group. Both reagents should be tried to determine optimum conditions. Neither reagent is always preferable and the use of one over the other will no doubt be dictated by individual circumstance, experience, and prejudice.

Either reagent converts the –NH–CO– residue into a –NH–CH_2– group. While this obviates the necessity for protecting nitrogen, careful consideration of other, reducible functional groups that may be present in the system is required.

It may be appropriate here to mention the possibility of reducing esters to ether linkages. Mechanistically the reduction of esters to ethers and amides to amines are closely related. The success of ester reduction has, to the Author's knowledge, been limited.

Considering the versatility of nitrogen-containing macrocycles, it would be worthwhile for additional studies of protecting groups to be undertaken. The allyl residue, for example, has been used to protect nitrogen and is reported to be cleaved using transition metal catalysis (Scheme 2.16). It was found in the Author's laboratory (unpublished) that hydrogenolysis of the allyl residue is only partially effective. Thus about half of the allyl crown was converted into the parent azacrown and the other half was converted into an *N*-propyl crown. If interest in macrocyclic chemistry continues to burgeon, there is no doubt that additional synthetic methods will be developed.

Which of these two approaches, or even others that can easily be imagined, is applied in the synthesis of azamacrocycles depends on several factors. These include;

The number of nitrogen atoms to be included
The ring size of the macrocycle desired
The presence of other functional groups, if any
Availability of the starting materials

Overall, then, the key difference between all-oxygen systems and azacrowns is that incorporation of nitrogen often requires use of a protecting group. Alternately, synthesis may be accomplished by a combination of lactam formation followed by reduction. Lactone formation for the synthesis of macrocyclic ethers by the conversion of esters directly to ethers is still a limited synthetic approach.

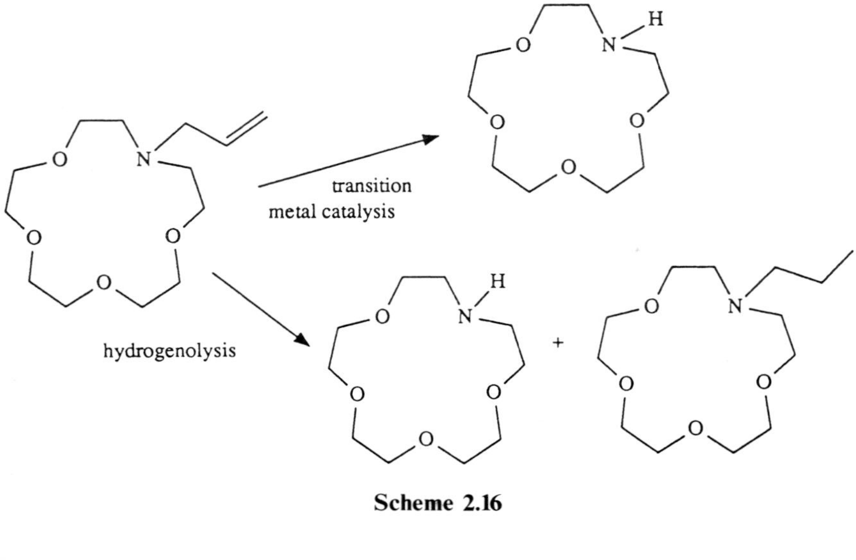

transition
metal catalysis

hydrogenolysis

Scheme 2.16

2.8.3 Direct Incorporation of Nitrogen

One way to completely circumvent the need for nitrogen protection is to use a primary amine as starting material and use both available positions in the cyclization. Dale and co-workers showed that protection was not required when both N–H bonds would be replaced by C–N bonds. Benzylamine has been used this way for the synthesis of its 12-crown-4 derivative and also for the direct synthesis of the diaza-18-crown-6 derivative. Dale and his co-workers converted benzylamine into *N*-benzylaza-12-crown-4. By using substituted benzylamine derivatives, this single step reaction afforded a range of aza-12-crown-4 derivatives for the first time. When R is benzyl in Scheme 2.17, it may be removed by hydrogenolysis to reveal aza-12-crown-4. This is the parent azacrown in the 12-membered ring series and may be used in the preparation of other derivatives.

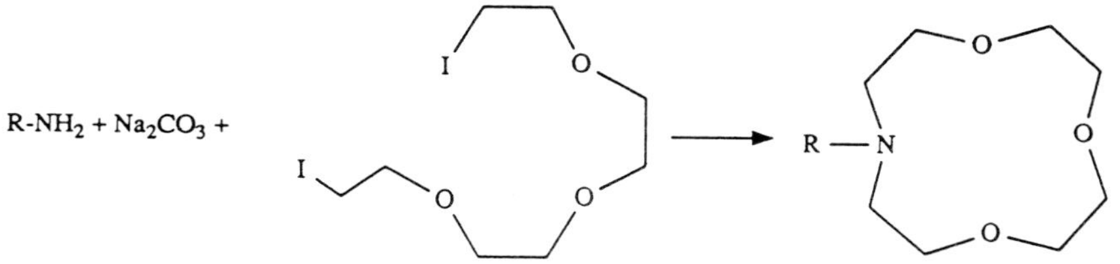

R-NH_2 + Na_2CO_3 + I–(CH₂CH₂O)₃–CH₂CH₂–I ⟶ R–N(aza-12-crown-4)

Scheme 2.17

Diaza-18-crown-6 derivatives may be prepared by a similar approach developed in the Author's laboratory by Vincent Gatto. If benzylamine was allowed to react with triethylene glycol diiodide in the presence of sodium carbonate and acetonitrile, a 2:2 reaction (Scheme 2.18) occurred rather than the 1:1 reaction shown in Scheme 2.17. The yields in the latter process were usually between 20% and 30%. Although this may seem low, we should not lose sight of two points. First, four bonds are formed in the process so a 30% yield means that each bond is formed in ~75% yield. Second, the reaction is conducted in reasonably concentrated solutions. Purification of these materials usually proved difficult unless the macrocycles crystallized as the sodium (from Na_2CO_3) complexes.

2.8.4 Syntheses of Lariat Ethers

The lariat ethers are represented by a large number of novel structures, prepared by many groups. The compounds studied in the Author's group were structures in which the macroring and sidearm could co-operate to bind a cation, in much the way a lasso is used to rope and tie a farm animal. The term is now applied to nearly any structure having a sidearm, whether or not binding is involved. Indeed, a number of such structures predate this non-systematic name but general methods for the synthesis of this family had not been developed. Examples of some *nitrogen-pivot*

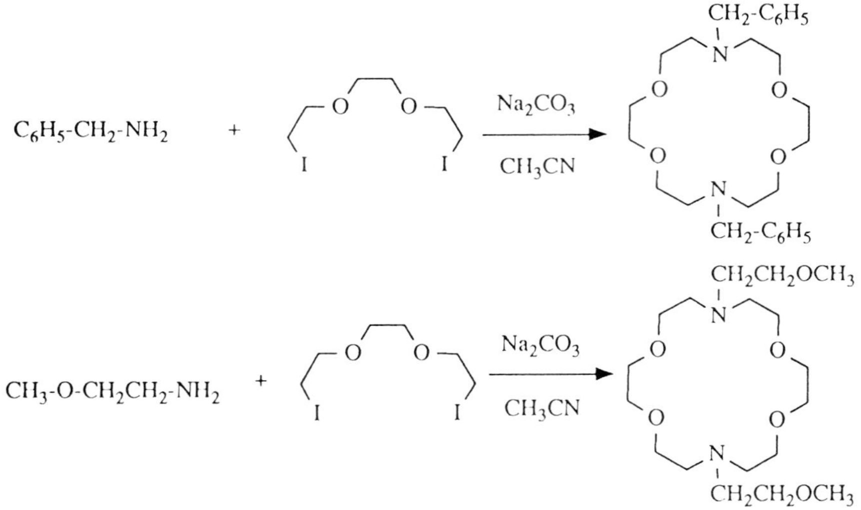

C_6H_5-CH_2-NH_2 + I–CH_2CH_2–O–CH_2CH_2–O–CH_2CH_2–I $\xrightarrow[CH_3CN]{Na_2CO_3}$ N,N'-bis(CH_2-C_6H_5) diaza-crown

CH_3-O-CH_2CH_2-NH_2 + I–CH_2CH_2–O–CH_2CH_2–O–CH_2CH_2–I $\xrightarrow[CH_3CN]{Na_2CO_3}$ N,N'-bis($CH_2CH_2OCH_3$) diaza-crown

Scheme 2.18

(sidearm attached to the macroring at N) lariat ethers, which would be called *podando–coronands* in the Weber–Vögtle nomenclature, are shown in structures **37–39**. Note that compound **39** was prepared by the Lehn group quite early.

The lariat ethers may be prepared in such a way that the incipient sidearm protects nitrogen during synthesis. Alternately, the readily available *N*-benzyl azacrown may be hydrogenolyzed to the parent and then alkylated (Scheme 2.19). When the former approach is used, diethanolamine is added to the incipient sidearm in an electrophilic form. Cyclization of the *N*-substituted diethanolamine derivative is then accomplished in the usual way (Scheme 2.19).

One interesting practical problem was encountered using these two approaches. When the sidearm was longer than about three ethyleneoxy units, the substituted diethanolamine derivative was itself a good com-

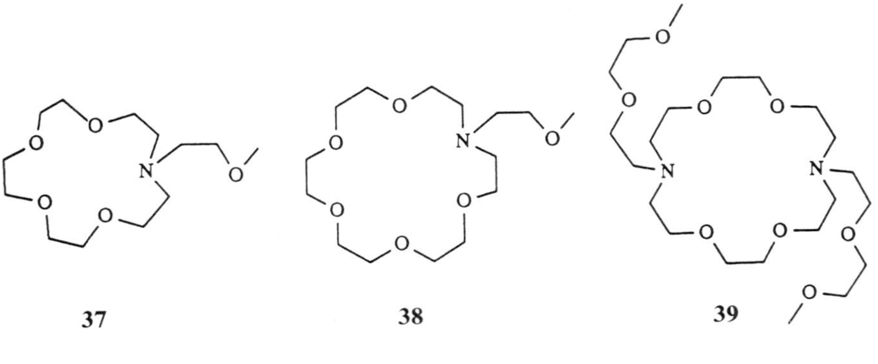

37 **38** **39**

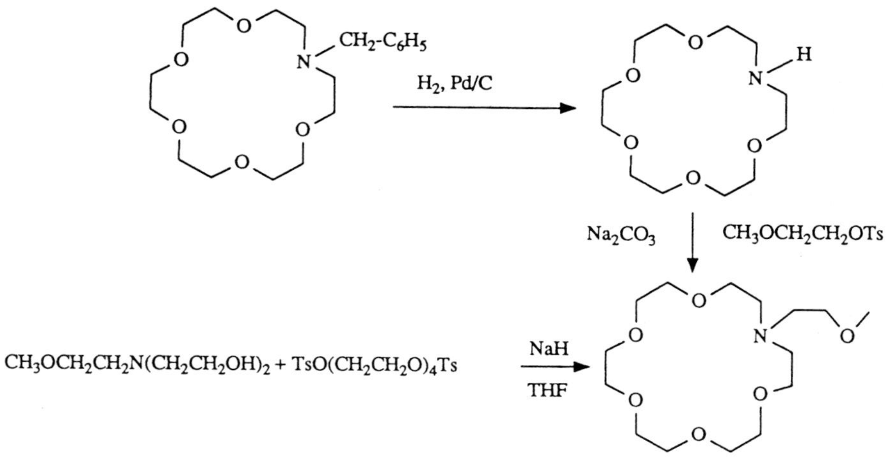

CH_2-C_6H_5 H_2, Pd/C H Na_2CO_3 $CH_3OCH_2CH_2OTs$

$CH_3OCH_2CH_2N(CH_2CH_2OH)_2$ + $TsO(CH_2CH_2O)_4Ts$ NaH THF

Scheme 2.19

plexing agent. It bound salts so tightly that purification became quite difficult. In such cases, the parent crown was prepared and alkylated rather than using the more direct approach.

2.8.5 Multi-armed Lariat Ethers

When two or more arms are attached to a macrocycle, the lariat or lasso image breaks down. Using the Latin word *bracchium* (arm), two armed systems were called **bib**racchial **l**ariat **e**thers, or **BiBLE**s. Likewise, three-armed systems are **trib**racchial **l**ariat **e**thers or **TriBLE**s. In principle, the nomenclature is systematically extensible to TetraBLEs, PentaBLEs, *etc.* Two lariat ethers and a BiBLE are illustrated in structures **37–39** of the previous section.

Two-armed, diaza-crowns have been studied by a number of workers over the years. These include van Zon, Reinhoudt, Lehn, Sutherland, Tsukube, Shinkai, Ricard, Vögtle, Bogatsky, Takagi, Kulstad and Malmstan, Cho and Chang, Bradshaw, Keana, the Author, and others too numerous to mention. Lehn's approach, noted above, to diaza-18-crown-6 involved *bis*(lactam) formation followed by reduction. Our single-step approach affords the 18-membered ring compounds very conveniently but the method is limited to that ring size.

An alternative to the latter approach was developed which afforded higher overall yields, albeit in a greater number of steps, but purification was generally easier. The strategy involved synthesis of the macrocycle subunit that contained both nitrogen atoms and the sidearms. This was then attached to a second chain to form the macrocycle. Since half the chain was already present and the sidearms were on nitrogen, no protecting group is required. This approach also proved more versatile in the sense that the second macroring-precursor unit could contain 2, 3, or more oxygens thus affording a range of diaza crown derivatives (Scheme 2.20).

It is probably worthwhile to mention at this stage that crown syntheses are often far more difficult than they appear on the surface to be. An excellent example of this is Lehn's original report of triaza-18-crown-6 (Scheme 2.21). This compound was a key precursor in his synthesis of the first spherand. The compound has considerable intrinsic interest but has been very little studied because the synthetic procedure Lehn published proved to be a reasonably complicated one. The Lehn procedure is fully reproducible but has generally been reported (largely by rumor) to be difficult to scale-up. A recently developed alternative to Lehn's synthesis is illustrated in Scheme 2.21.

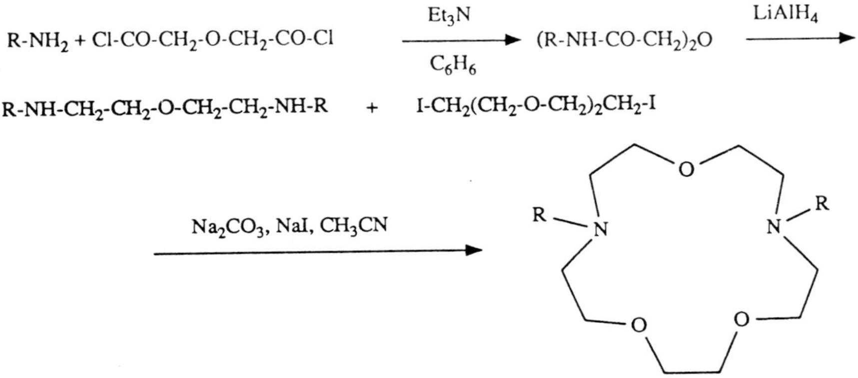

$$R\text{-}NH_2 + Cl\text{-}CO\text{-}CH_2\text{-}O\text{-}CH_2\text{-}CO\text{-}Cl \xrightarrow[C_6H_6]{Et_3N} (R\text{-}NH\text{-}CO\text{-}CH_2)_2O \xrightarrow{LiAlH_4}$$

$$R\text{-}NH\text{-}CH_2\text{-}CH_2\text{-}O\text{-}CH_2\text{-}CH_2\text{-}NH\text{-}R \quad + \quad I\text{-}CH_2(CH_2\text{-}O\text{-}CH_2)_2CH_2\text{-}I$$

$$\xrightarrow{Na_2CO_3,\ NaI,\ CH_3CN}$$

Scheme 2.20

Synthesis of 4,10,16-Triaza-18-crown-6

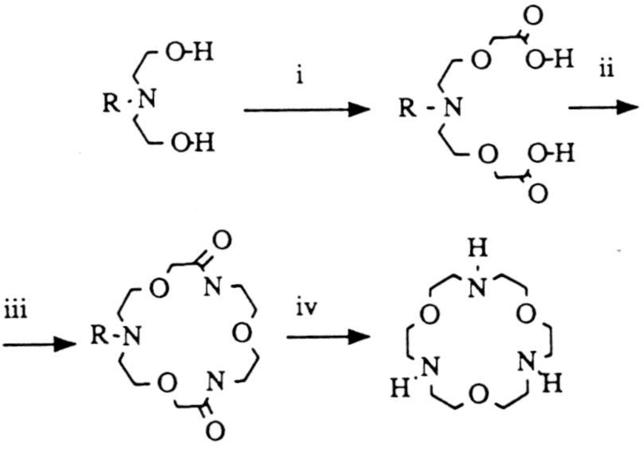

R = Tosyl. i. $ClCH_2COOH$, NaH, THF, 25 °C, 24 h, 75%; ii. ClCOCOCl, C_6H_6 and catalytic C_5H_5N, quantitative; iii. $H_2NCH_2CH_2OCH_2CH_2NH_2$ in C_6H_6, high dilution, product used directly in step iv; iv: $LiAlH_4$, THF, reflux 36 h, 65%.

Scheme 2.21

2.8.6 Heteroaromatic Sub-units

The incorporation of heteroaromatic sub-units into macrocycles is not a new concept. Nature has done this at least since the first plants grew. Chlorophyll and hemoglobin are two of the best known examples of natural, cation-binding macrocycles. The porphyrin system contains four pyrrole rings and is closely related to the macrocycle containing four furan subunits illustrated earlier in this chapter (see Scheme 2.14). The nitrogen atoms in a porphyrin are clustered in a square plane and are sufficiently far apart to accommodate relatively small metals such as magnesium or transition metals. Iron (II), for example, is nicely accommodated in the porphyrin ring of hemoglobin just as Mg^{2+} fits in the chlorophyll porphyrin ring. The general structure of chlorophyll is in structure **40**. When R is methyl, the compound is chlorophyll and when R = formyl (-CH=O), it is chlorophyll b.

Almost as soon as crown ethers were discovered, chemists began to incorporate heteroaromatic sub-units into the crown ether ring. Incorporation of pyridine, a study conducted in Cram's laboratory, was one of the earliest cases reported. A 2,6-dimethylpyridine unit can substitute nicely for a diethyleneoxy unit. The basic electron pair on nitrogen is focused to the center of the macroring. The synthesis of this compound actually proved to be fairly straightforward. Tetraethylene glycol ditosylate was treated with 2,6-*bis*(hydroxymethyl)pyridine (**41**) under basic conditions (Scheme 2.22). The product, a white solid, was obtained in about 30% yield.

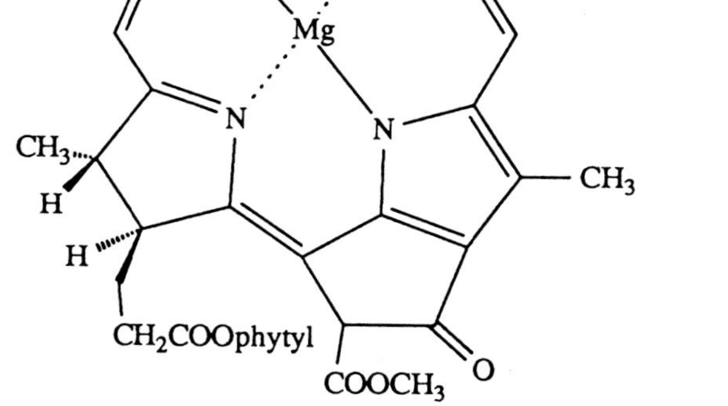

40

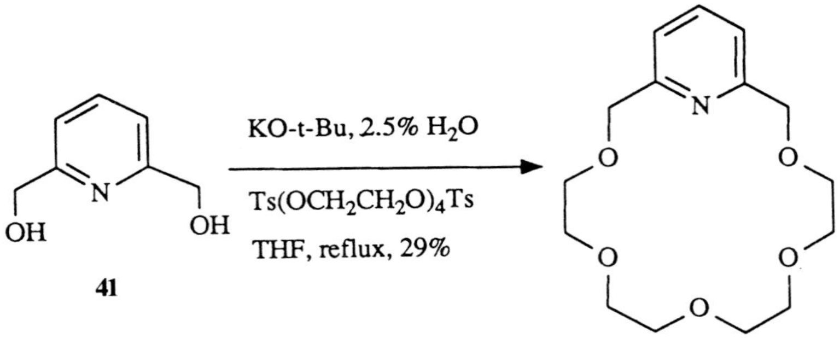

KO-t-Bu, 2.5% H_2O

$Ts(OCH_2CH_2O)_4Ts$

THF, reflux, 29%

OH OH

41

Scheme 2.22

A similar structure was prepared by Bradshaw and co-workers whose interest at the time focused on macrocyclic lactones. One of their key findings was that the reactions of oligo(ethylene glycol)s and carboxylic acid halides afforded a mixture of polyesters which appeared to be largely linear. Pyrolytic distillation of this mixture afforded the macrocycle as a 'monomer' in good yield. Thus, the analog of the 18-membered ring pyridine compound shown in Scheme 2.22 was prepared in 77% yield (Scheme 2.23). The binding properties of this material compared favorably with the pyridine containing polyether. This is not surprising since the carbonyl groups are directed outward and should not interfere with (or enhance, for that matter) the cation binding ability. Bradshaw has reviewed the area of macrocyclic ester syntheses and the reader is referred to those reviews for a more detailed description of this approach.

Newkome and co-workers were particularly active in the pyridine macrocycle area. Although a large number of compounds were prepared by this group, by far the most interesting of them was the analog of 18-crown-6 which contained six pyridine units. Newkome referred to this

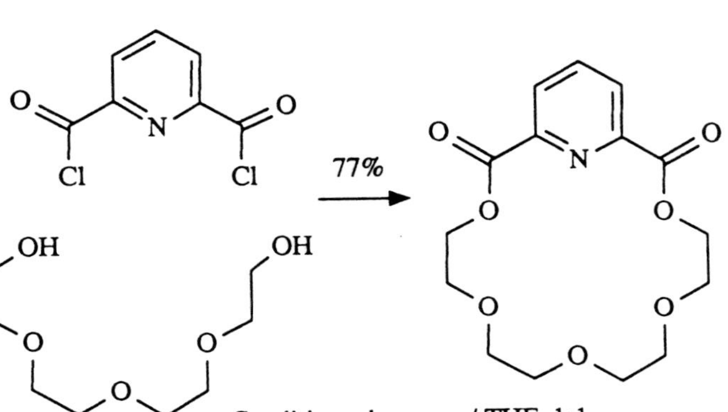

77%

Conditions: benzene / THF, 1:1

Scheme 2.23

molecule, which was also a target in Toner's program (at Kodak), as 'sexi-pyridine'. The molecule is a direct analog of 18-crown-6 in which all of the carbon atoms are sp^2 hybridized and all of the donor groups are nitrogen atoms. It is presumably the difficulty of synthesis and the small quantities thus available that have prevented complexation studies from being conducted. More recently, Bell has extended this work and shown that sexi-pyridine derivatives are capable of complexing certain metal ions very strongly. The unsubstituted parent compound is illustrated in structure **42**.

Another contribution made by Newkome and co-workers involved direct nucleophilic substitution reactions on substituted pyridine derivatives. An example of this approach is illustrated in the cryptand discussion following (Section 2.10). We show two of the structures, **43** and **44**, prepared by this method.

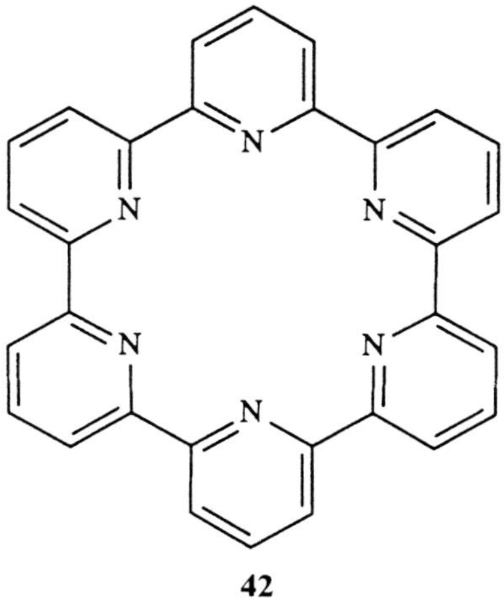

N N N N N N

42

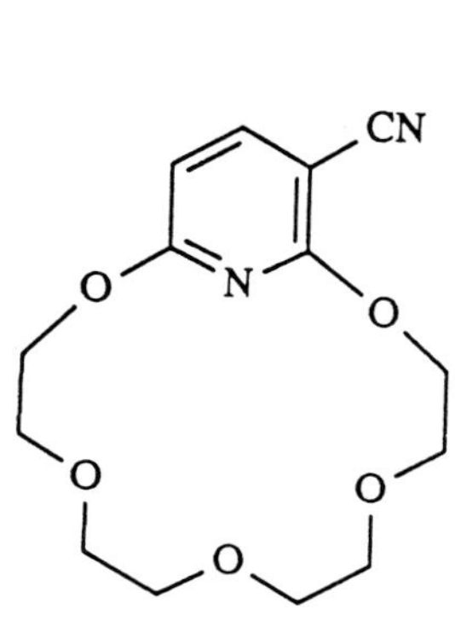

CN N O O O O O

43

N N O O O O O O

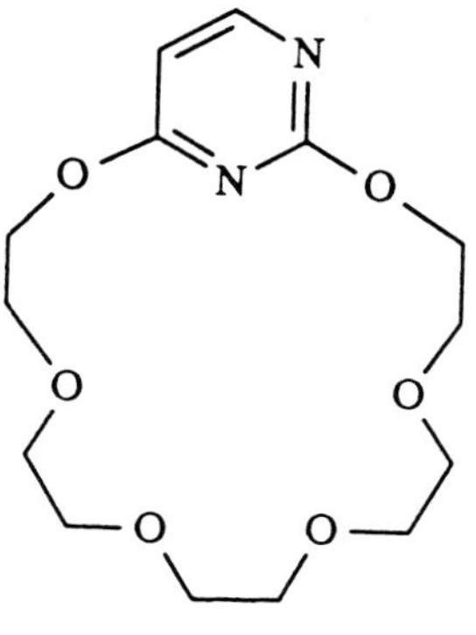

44

Kellogg and his co-workers exploited macrocyclic pyridine-containing compounds as models for the biological redox agent, NAD. The *N*-methylpyridinium system is electron deficient and capable of receiving a hydride from a cysteine-bound 3-phosphoglyceraldehyde. The pyridine salt is thus converted into a dihydropyridine derivative. The sequence is believed to occur as shown at the left, in Scheme 2.24. One of Kellogg's crown mimics is shown at the right. Note that this structure incorporates chirality as well as the pyridine nucleus.

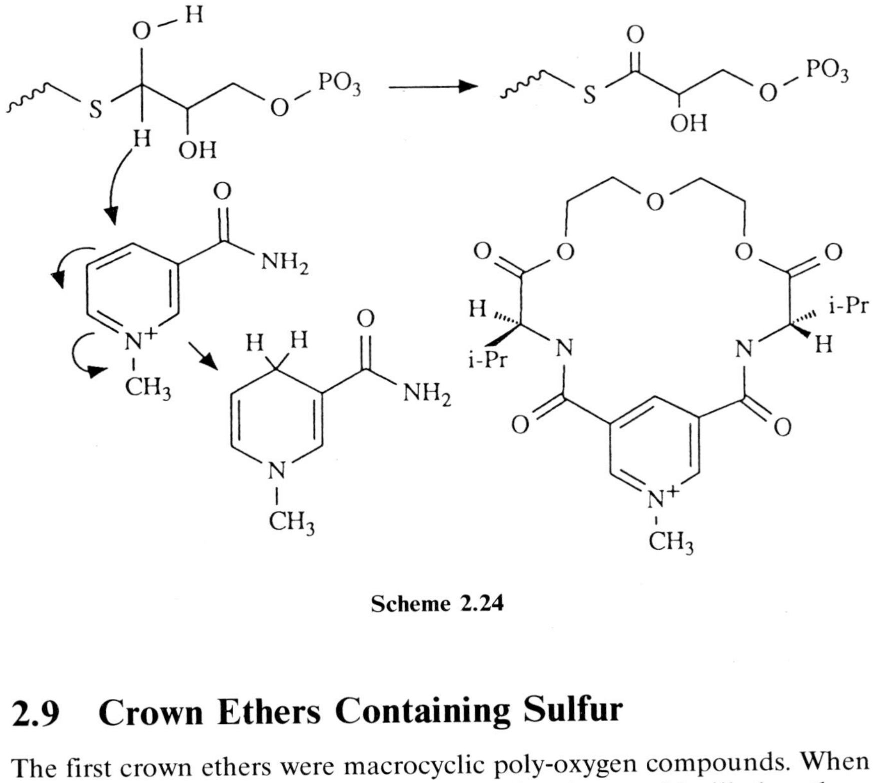

Scheme 2.24

2.9 Crown Ethers Containing Sulfur

The first crown ethers were macrocyclic poly-oxygen compounds. When one or two oxygen atoms are added, most chemists would still place these structures within the organic domain. A macrocycle having one or two nitrogen atoms but having sulfur atoms as the only other donors would probably be considered to be in the inorganic domain. This distinction is certainly artificial. It is the case, however, that compounds containing a high proportion of sulfur and nitrogen are usually found in the smaller macrorings and these are usually more appropriate for complexing transition rather than alkali metals.

Early in the development of crown ether systems, the group of Izatt, Christensen, Bradshaw and their co-workers was curious about how substitution of sulfur for oxygen affected the cation binding properties of macrocyclic polyethers. It was anticipated that if one or more sulfur atoms replaced oxygen, silver or transition metals would be favored. Indeed, this expectation proved to be correct.

The synthesis of sulfur-containing macrocycles is generally a straightforward process. The difficulty is that the starting materials are often hard to handle. One can imagine incorporating sulfur into a macroring simply by substituting $ClCH_2CH_2SCH_2CH_2Cl$ for $ClCH_2CH_2OCH_2CH_2Cl$. The former is very reactive, however, and is both a lachrymator and vesicant. The synthesis of sulfur-containing compounds by this route is therefore attended with some difficulty. Despite these problems, a number of sulfur-containing and mixed sulfur–nitrogen compounds have been prepared. Examples of an aliphatic per-sulfur derivative and a mixed S–N compound prepared by aromatic substitution are shown in Scheme 2.25 and Scheme 2.26.

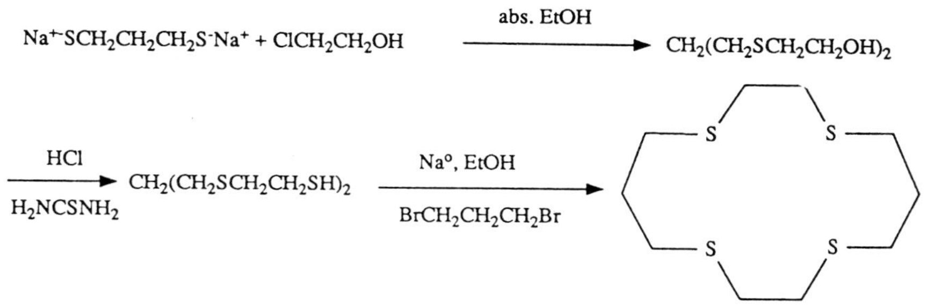

$Na^{+-}SCH_2CH_2CH_2S^-Na^+ + ClCH_2CH_2OH \xrightarrow{\text{abs. EtOH}} CH_2(CH_2SCH_2CH_2OH)_2$

$\xrightarrow[H_2NCSNH_2]{HCl} CH_2(CH_2SCH_2CH_2SH)_2 \xrightarrow[BrCH_2CH_2CH_2Br]{Na^o,\ EtOH}$ S S S S

Scheme 2.25

N Cl N Cl $\xrightarrow[(CH_2SH)_2]{NaH}$ N S N Cl S N Cl N

$\xrightarrow[O(CH_2CH_2OH)_2]{NaH}$ N S N O O O S N N

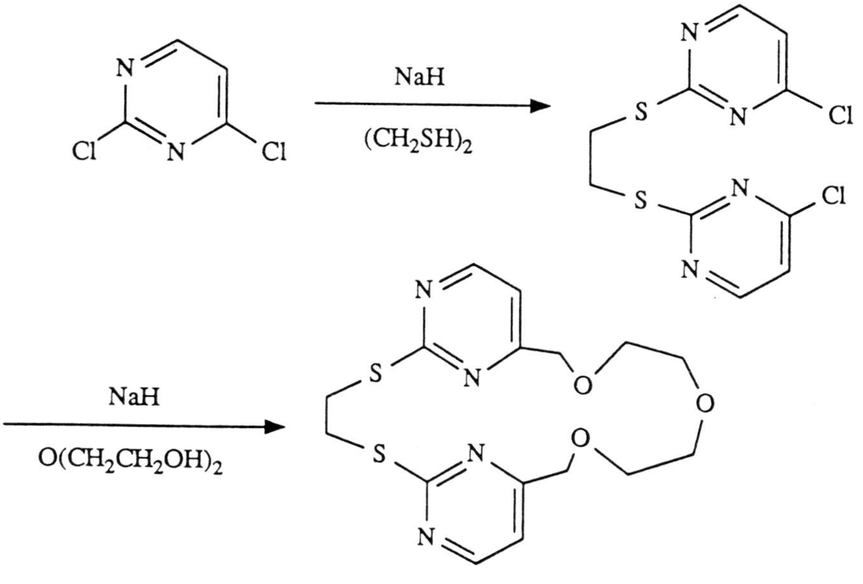

Scheme 2.26

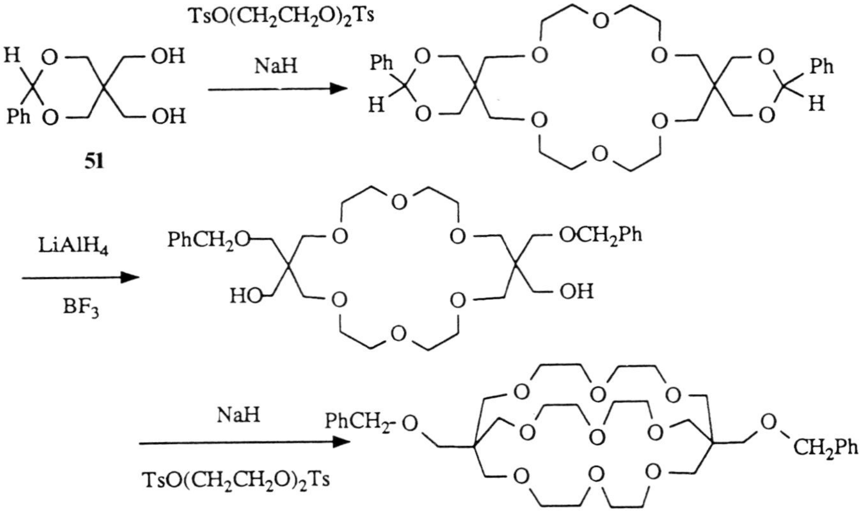

$TsO(CH_2CH_2O)_2Ts$

NaH

51

$LiAlH_4$

BF_3

NaH

$TsO(CH_2CH_2O)_2Ts$

Scheme 2.29

2.11 Spherands, Cavitands, and Carcerands

Among the most complex cryptands are the spherands. Two structures are shown (**52** and **53**). The one at the left (**52**) was Lehn's first spherand and the one at the right is the product of Cram's imagination. Lehn's

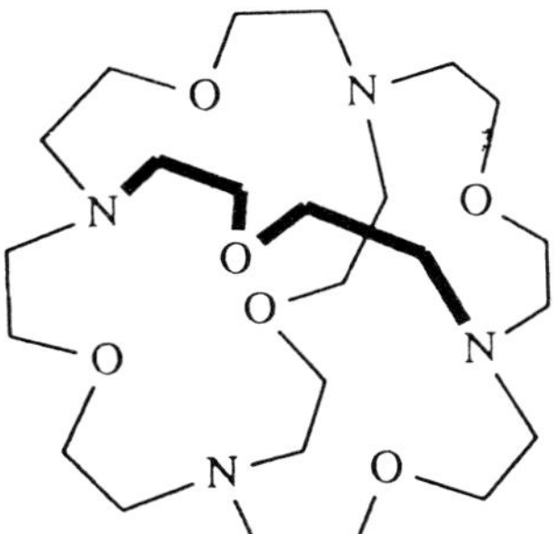

52 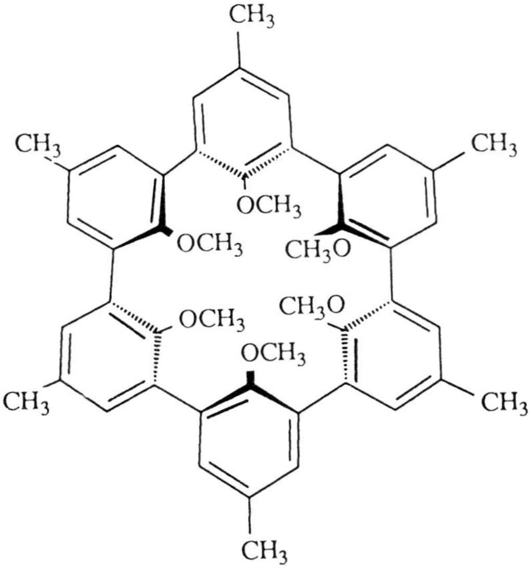 **53**

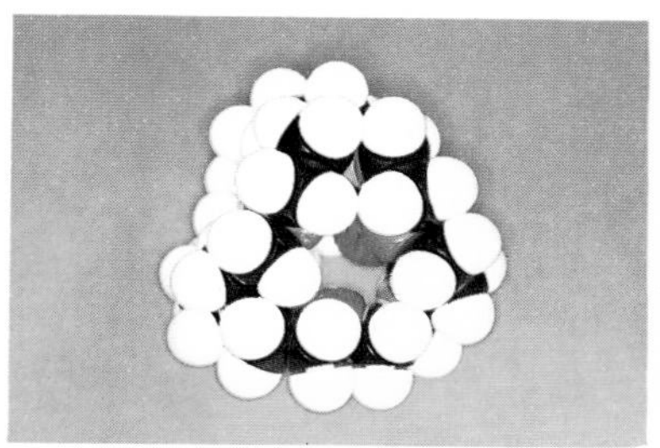

CPK atomic models (two views) of Lehn's spherand in the absence of a guest

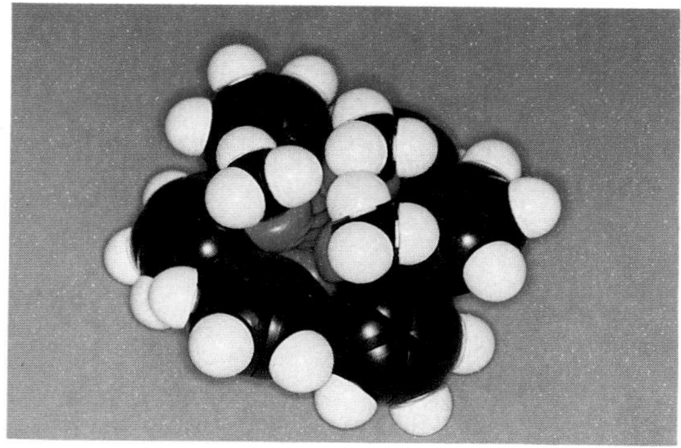

CPK atomic model of Cram's spherand (note absence of *para*-methyl groups)

spherand was prepared in a fashion largely similar to that used for the cryptands except that triaza-18-crown-6 rather than diaza-18-crown-6 was the key intermediate.

The synthesis of the hexa-phenyl spherand **54** is quite different from most of the syntheses discussed heretofore (Scheme 2.30). This is because the compound is constructed exclusively from aromatic rings. This approach confers upon the host (ligand) considerable rigidity and organization. Nevertheless, the oxygen atoms are in a structural array similar to that of 18-crown-6, although the O–O distances are obviously different.

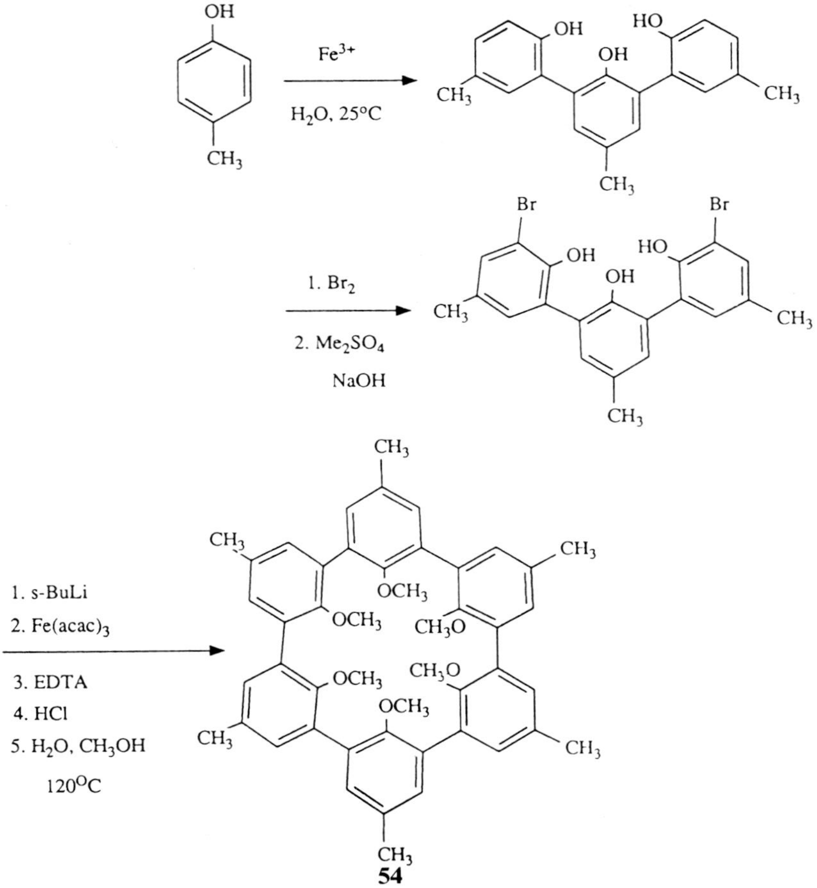

OH

Fe^{3+}

H_2O, 25°C

CH_3

OH HO OH CH_3 CH_3 CH_3

Br Br OH HO OH

1. Br_2
2. Me_2SO_4 NaOH

CH_3 CH_3 CH_3

1. s-BuLi
2. $Fe(acac)_3$
3. EDTA
4. HCl
5. H_2O, CH_3OH 120°C

CH_3 CH_3 CH_3 OCH_3 OCH_3 CH_3O OCH_3 CH_3O OCH_3 CH_3 CH_3 CH_3

54

Scheme 2.30

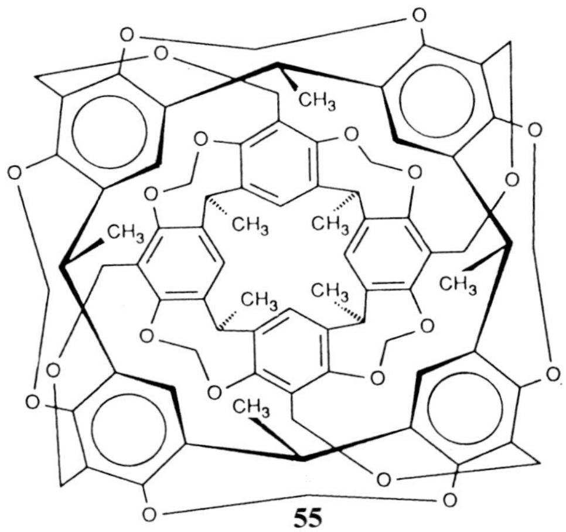

55

It is interesting to note in passing that there is substantial similarity between the Cram spherands, carcerands (**55**) and the long-known and recently popular molecules now called *calixarenes*. The latter are produced by condensation of a phenol with an aldehyde. The cyclotriveratrylenes are also accessible by similar technology. The compound now called calix[6]arene is illustrated **56**.

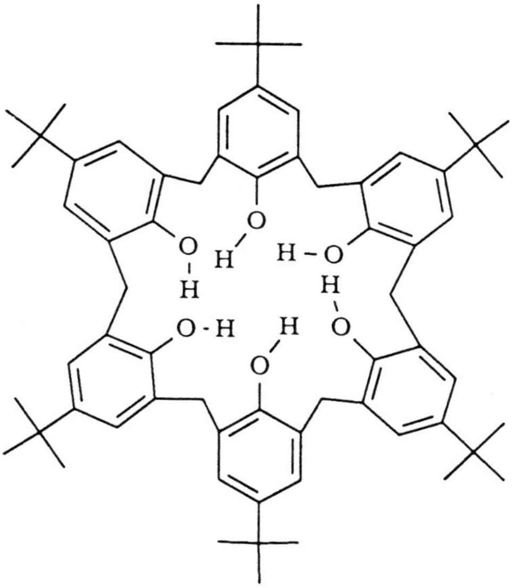

56

2.12 Summary

The syntheses of crown ethers and cryptands involve all of the difficulties and subtleties encountered in natural product synthesis. Strategies must take account of ring size, functional group compatibility, and often, stereochemistry. An important difference between most natural product syntheses and those undertaken by workers in the macrocycle area is that specific properties rather than a specific structure may be the target. Thus the synthesis is complicated by anticipating the physical chemistry of the product as well as considering the tactical difficulties. The synthetic chemist unfamiliar with the problems of macrocycle synthesis may well be surprised to discover that these highly symmetrical and often simple-looking structures present complex synthetic problems. The reward, of course, is to prepare a hitherto unknown structure and to discover that this 'unnatural product' has the intended chemical properties.

CHAPTER 3

Complexation by Crowns and Cryptands

3.1 Introduction

Macrocyclic compounds are intrinsically beautiful and intriguing structures. Their sizes, shapes, and surfaces have fascinated generations of chemists. The crowns and cryptands have profoundly expanded previously known structural possibilities but it is the complexation process that has made all of these compounds so important. Simple aliphatic macrocycles such as cyclotetradecane or even adamantane (Figure 3.1) have interesting structural characteristics, but the absence of donor groups prevents them from direct interaction with substrates except by simple, van der Waals (lipophilic–lipophilic or hydrophobic–hydrophobic) interactions.

In his first paper on crown ethers, Pedersen recognized the ability of crown ethers to complex cations. Indeed, it was this very property of which he was in search. He had intended to prepare Ca^{2+}-complexing agents having the structure **1** as shown.

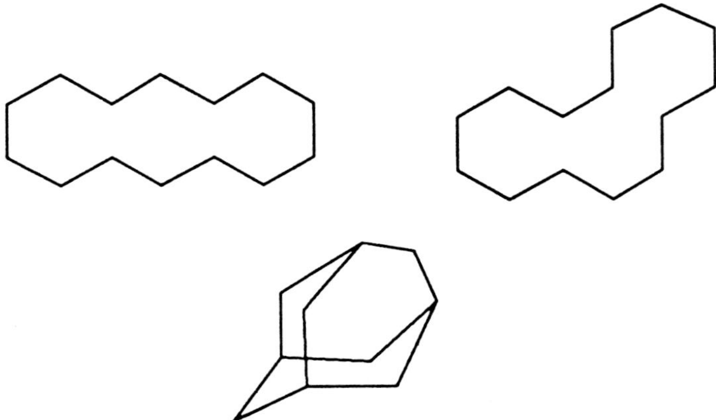

Figure 3.1 *Flexible cyclotetradecane shown (above) in two of its many possible conformations and rigid ('diamond-lattice') adamantane*

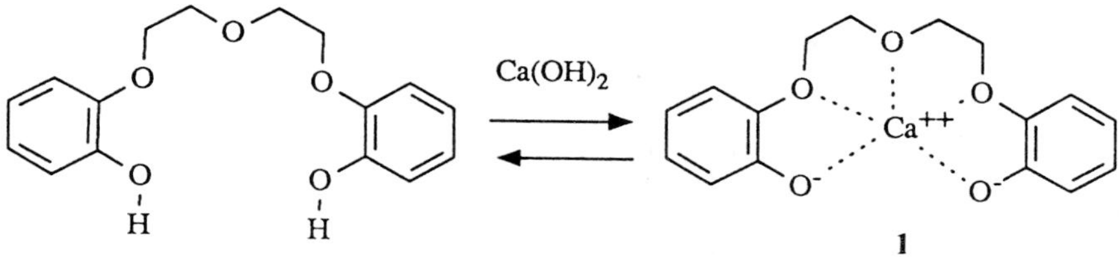

Pedersen prepared these compounds precisely because he expected the free phenolic residues to dissociate under basic conditions. A doubly-negative, bidentate ligand would result and this should have a high affinity for calcium or other divalent species. Indeed, Pedersen prepared and patented these compounds. It was the by-product of this synthesis that led to the crown ethers.

At the time Pedersen undertook his work, neutral complexing agents for transition and coinage metals were already well known, but complexation of alkali metals by neutral species was another matter entirely. Pedersen examined his new complexing agents using ultraviolet spectroscopy (sensitive to the aromatic ring chromophore). It had been no surprise when the spectrum changed as the phenol(s) ionized; a spectral change for the neutral compound in the presence of metal salts was remarkable. The reaction Pedersen had observed is shown in Scheme 3.1.

The complexation process has now been studied extensively and thoroughly. Numerous approaches have been used over the years but a few have emerged as most convenient or most generally useful. Two of these were pioneered by Pedersen and his colleague at Dupont, Frensdorff. These methods are the extraction technique and ion selective electrode techniques.

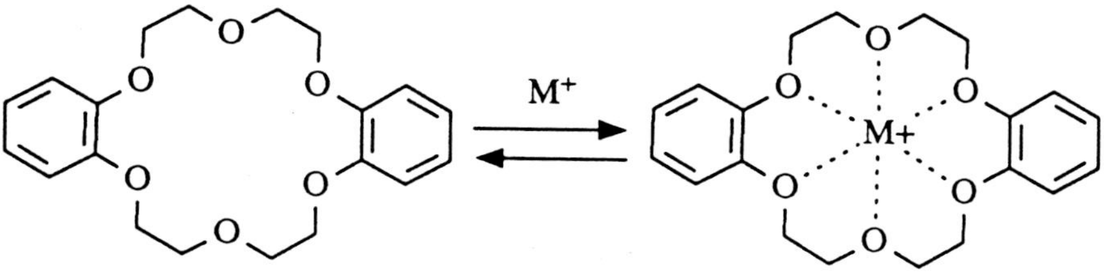

Scheme 3.1

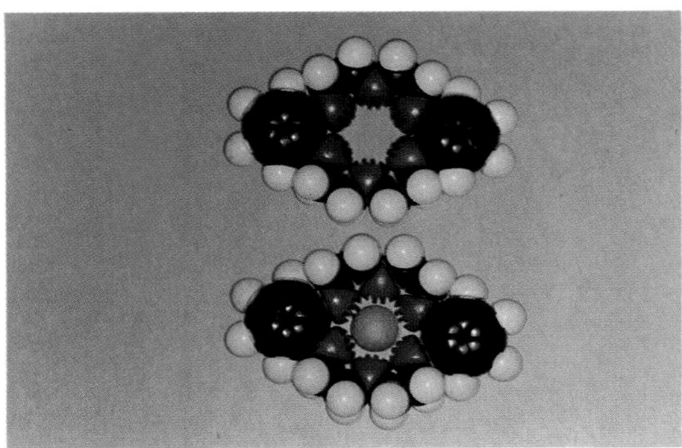

CPK atomic model of dibenzo-18-crown-6 in the binding conformation in the presence and absence of K^+

3.2 The Extraction Technique

Assessment of cation binding strengths and dynamics are of central importance to an understanding of ionophore properties. One of the earliest ways of assessing cation binding strengths utilizes a method widely known as the 'extraction technique'. This approach uses the fact that most crown ethers are insoluble or nearly insoluble in water (especially salt-containing water) and that most metal salts are insoluble in organic solvents. Thus, dibenzo-18-crown-6 dissolves readily in chloroform or dichloromethane but is insoluble in water. Likewise, NaCl is freely soluble in water but insoluble or nearly so in most organic solvents.

Equal volumes of chloroform and water may be mixed by shaking but will eventually return to two separate phases. When dibenzo-18-crown-6 is added to the mixture, it partitions mostly into the organic phase. Likewise, a salt such as sodium picrate dissolves in the aqueous phase (Figure 3.2). Picric acid, because there are three nitro groups present on the aromatic ring, is yellow and it can readily be detected by eye or quantitatively by ultraviolet spectrometry. A mixture of chloroform, water, and sodium picrate in a vessel exhibits two phases in which the top, aqueous phase is yellow. Even if shaken, the chloroform phase remains colorless. When crown ether is added, it dissolves in the chloroform phase. The ligand's complexation behavior causes it to bind sodium cation. Since the cation cannot be extracted into chloroform in the absence of some anion, picrate partitions into the organic phase as well. If all of the sodium picrate is extracted into the crown/chloroform phase, the aqueous phase will become colorless. The extent to which extraction occurs can be quantitated by colorimetric methods using Beer's law.

In a typical experiment, one equivalent of a crown ether and one equi-

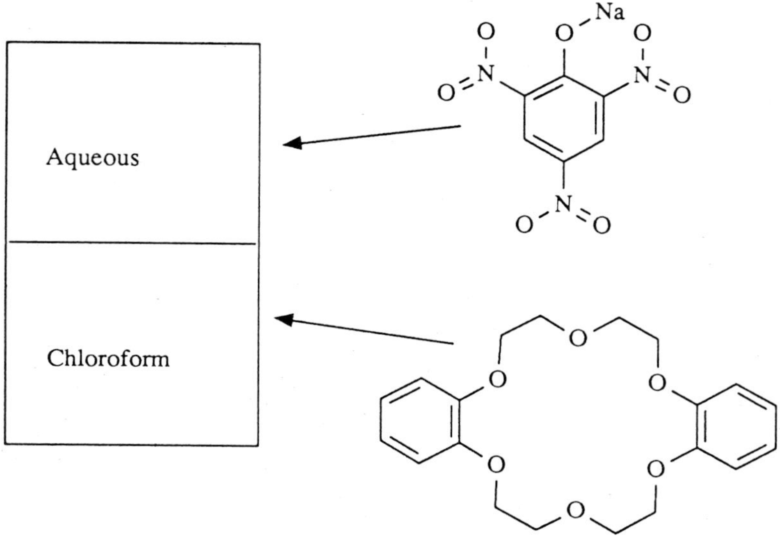

Figure 3.2 *Partitioning of crown and sodium picrate between water and chloroform solution*

valent of sodium picrate are partitioned between chloroform and water (Figure 3.2). Let us assume that 50% of the available salt (as judged by a quantitatively determined color change) is extracted into chloroform. The extraction constant would be 50%. If, under identical conditions, 70% of potassium picrate was extracted, the *extraction constant* would be 70% and the crown ether would be pronounced selective for potassium cation. Indeed, as described here, such a conclusion would be valid for this system.

Let us now assume that only half an equivalent of salt was used in each case and, hypothetically of course, that all of the available salt was extracted in each case. The extraction constants would be 100% in both cases and, by the definition:

$$\text{Selectivity} = \text{extraction constant } 1 \div \text{extraction constant } 2 = 100\%/100\% = 1$$

The crown, according to this experiment, exhibits no selectivity. The problem here is obvious. One must construct the experiment in such a way that meaningful information will be obtained. Experimental design is always a challenge in physical organic chemistry but it is critically important in a case such as this one where design of ligands might be based on the mistaken belief that a crown ether lacked selectivity simply because the data were determined under inappropriate conditions.

Are there correct conditions for determination of extraction constants? Indeed, if extraction constants are so dependent on experimental conditions, can they be called 'constants' at all? The answer is that the data reflect the conditions used to determine them and must be compared only with data obtained under identical conditions. Whether or not data determined in such a way are reliable indicators of selectivity depends on the point of view. Complexation strength *vis-à-vis* the phase transfer catalysis method might be correctly indicated since that process, like the extraction technique, involves partition between two phases.

It is relatively difficult (and not particularly safe) to prepare anhydrous picrate salts. It is also not a particularly efficient approach if one is then to dissolve the salts in water. Thus, a favored technique is to dissolve picric acid in aqueous NaOH or KOH solution to afford the salt. If exactly equal amounts of acid and base are present and neutralization is complete, then a pure salt solution exists. If an excess of base is used, as is often the case, then the same amount of metal picrate may be present in solution, but the additional base will affect the medium's ionic strength and this, in turn, will affect the extent of extraction. The identity of the two solvents chosen, the solvent volumes, temperature, ionic strength, and other variables affect the extraction constants. If, however, all conditions are reproduced exactly in another set of experiments, the data are regarded as comparable. Unfortunately, extraction constant data are often reported in the absence of comparable data for standard compounds such as dibenzo-18-crown-6. Thus, internal comparisons are possible among the structures surveyed, but a calibration point is often lacking.

In the early stage of our study of lariat ether compounds, we prepared several derivatives of 15-crown-5 having substituted aryloxymethyl sidearms. We wished to assess whether or not the sidearm donor group participated in cation complexation. An examination of space-filling, CPK molecular models, suggested that when the sidearm donor was *ortho*, participation was possible but when the sidearm donor was *para*, it was too remote to participate. Extraction constants (dichloromethane/water) were determined for the compounds **2** and **3**.

It appears from the data that nearly all of these 15-membered ring compounds are selective for Na^+ over K^+. Such a selectivity has often been rationalized as a 'hole size' effect, *i.e.* a correspondence between cation

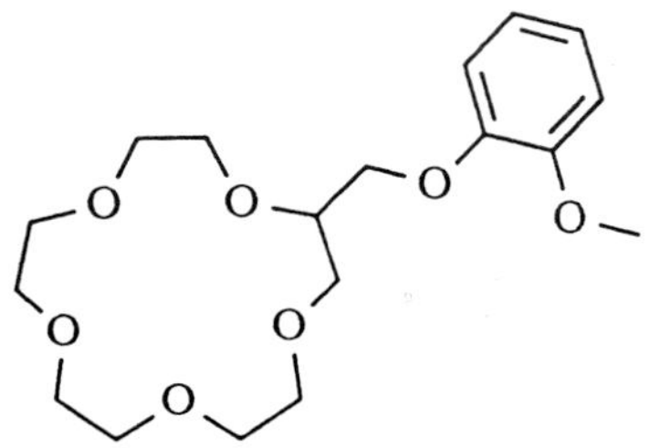

2

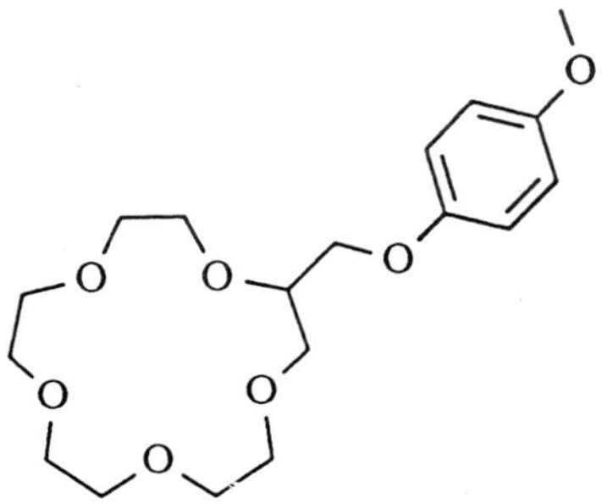

3

Table 3.1 *Extraction constant data for carbon-pivot lariat ethers*

Sidearm on 15-crown-5	*Extraction const.* Na^+	K^+	*Selectivity* Na^+/K^+
H(15-crown-5)	7.6	5.7	1.33
CH_2OCH_3	5.1	3.3	1.55
2-phenoxymethyl	4.0	4.3	0.93
2-(*ortho*-methoxyphenoxymethyl)	15.7	10.2	1.54
2-(*para*-methoxyphenoxymethyl)	6.4	10.7	0.60

size and macrocycle hole or cavity size. Further, it appears that Na^+ binding is dramatically increased by the lariat ether sidearm. It is less clear, however, why the *para* compound selects K^+ over Na^+ and why K^+ is bound equally well by both the *ortho* and *para* isomers. Keeping in mind that the extraction constant technique requires partitioning between an organic and an aqueous phase, it might be that binding is enhanced by increasing lipophilicity rather than appropriate placement of the sidearm donor group.

Another troubling issue is selectivity. Why should the selectivities for the *ortho* derivative and the methoxymethyl derivative be identical when their binding strengths are very different? Why should the K^+ binding strengths be identical for the *ortho* and *para* isomers when their selectivities are inverted? No simple answers to these questions are yet available. It is interesting to note, however, that entirely different results are obtained when homogeneous cation binding strengths are determined. The latter technique is discussed below.

The extraction technique is a fast and simple method for assessing binding strengths. It has even been used in a number of cases to estimate free energies (ΔG) of binding. The technique is internally consistent and data are comparable to those obtained for other systems when identical conditions are used. Unless identical experimental conditions are used, however, conclusions drawn from two sets of extraction constants may be suspect. Further, it appears that the extraction technique is sensitive to hydrophilic/hydrophobic balance and that this may affect conclusions. The technique should be used with attention to detail and with due caution.

In an early study of 18-membered ring compounds binding *tert*-butylammonium cation, Cram and co-workers developed a different type of extraction procedure. In this case, the salt (guest) and ligand (host) were partitioned between $CDCl_3$ and D_2O. Partitioning between the organic and aqueous phases was assessed by ^1H NMR. The partition equilibrium was:

$$[\text{host}] + [(CH_3)_3CNH_3^+SCN^-] \underset{CDCl_3}{\overset{K_a}{\rightleftharpoons}} [(CH_3)_3CNH_3^+ \cdot \text{host} \cdot SCN^-]$$

The binding constant was assessed according to the formula

$$K = \frac{[BX]_{D_2O}{}^2 R}{[BX]_{CDCl_3}(1-R)\{[BX]_i-[H]_i R(V_{CDCl_3}/V_{D_2O})\}^2}$$

in which $[BX]_{D_2O}$ and $[BX]_{CDCl_3}$ were equilibrium concentrations of salt in the absence of host, R is the ratio of concentrations of guest to host in $CDCl_3$ at equilibrium, $[BX]_i$ is the initial salt concentration in D_2O, $[H]_i$ is the initial host concentration in $CDCl_3$, and V_{D_2O} and V_{CDCl_3} are the volumes of D_2O and $CDCl_3$. By using this technique, Cram was able to assess the effect of structural variations within a fairly similar group of compounds. A few of the compounds studied (at 24 °C) and reported are shown (**4–9**) along with their equilibrium distribution constants.

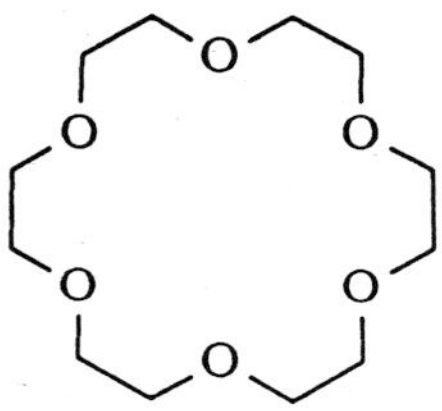

7.5x10^5

4

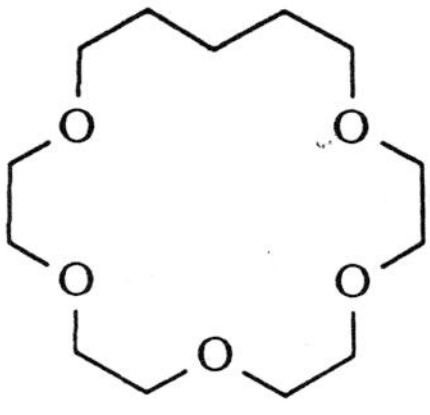

5.0x10^2

5

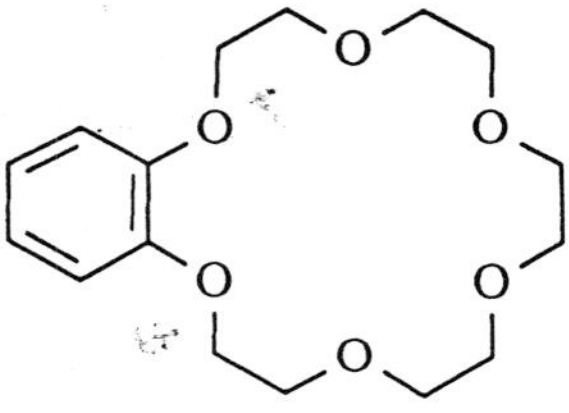

1.3x10^4

6

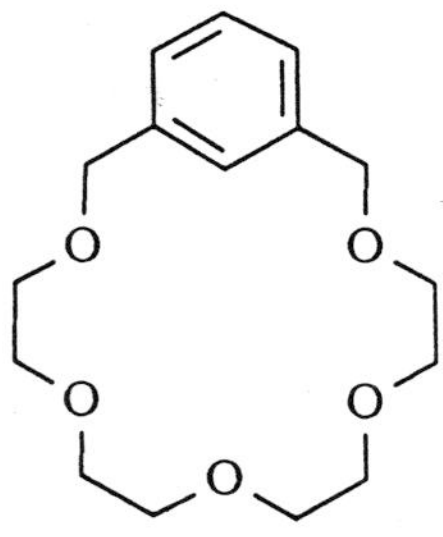

1.5x10^3

7

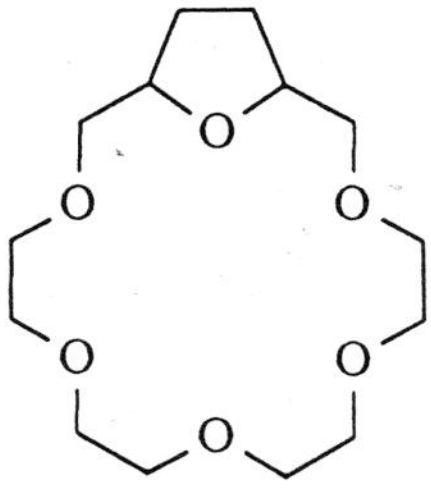

1.1x10^6

8

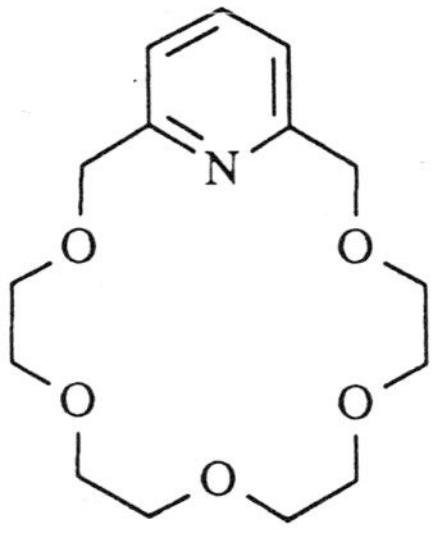

1.4x10^6

9

First, a comment about this method is in order. Few convenient alternatives to this method exist that might be used to assess the binding of alkylammonium cations. The ammonium cation itself, NH_4^+ can be studied by using ion selective electrode methods but this method has not, to the Author's knowledge, been applied to alkyl or arylammonium cations. Calorimetry could be and has been used in some cases, but the binding must be within the instrumental window of the method. Homogeneous cation binding constants might be accessible by monitoring NMR spectral changes as a function of host/guest ratios or by using conductometric techniques. The range of binding presented in these structures (in this solvent, at the specified temperature, *etc.*) is about 10^4. In short, one can hardly criticize a method if information is obtained that might be difficult to obtain or unavailable otherwise. In the Author's opinion, use of the extraction technique is more valuable in a case such as this one than for binding of metallic cations for which other methods that afford directly comparable data are available.

Second, the trends exhibited by these binding studies help to understand how cation binding strengths vary as a result of structural variation. For example, changing –O– to $-CH_2-$ reduces the association constant by more than a thousand-fold. This is due to the loss of a donor atom, the conformational alterations attending the change from $-CH_2-CH_2-\boldsymbol{CH_2}-CH_2-CH_2-$ (pentylene) to $-CH_2-CH_2-\boldsymbol{O}-CH_2-CH_2-$ within the macroring structure and more subtle features such as changes in dipoles, polarity, and lipophilicity. There is less conformational mobility when the *meta*-benzo unit replaces the pentylene unit. There is also an increase in lipophilicity and both undoubtedly contribute to the increased binding strength. When a donor atom is reinstated into this aromatic unit (*i.e.* when pyrido replaces benzo), the host is an even better binder than 18-crown-6. It is interesting to note that the difference in donor atom hybridization and base strength is irrelevant for the tetrahydrofuranyl derivative compared to the pyrido compound. Finally, the influence of sp^2 hybridization compared to sp^3 hybridization is also apparent, although lipophilicity and conformational rigidity must surely play a role.

3.3 Homogeneous Cation Binding Constants

Homogeneous cation binding constants, sometimes called *stability constants*, are nothing but the equilibrium constants for a reaction such as that shown in Figure 3.3. When a macrocycle reacts with a cation (the anion is omitted in Figure 3.3), a complex is formed. The position of this equilibrium is usually given the designation K_S. As with any equilibrium constant, it is the ratio of the rates for the forward (k_f) and reverse (k_r) reactions. The forward reaction rate is also sometimes referred to as $k_{complex}$ or simply as k_1. The reverse reaction is correspondingly referred to as $k_{release}$ or $k_{decomplex}$ or simply as k_{-1}. The rates for many complexation reactions have been determined and are discussed below. For the moment, let us consider cation binding strengths.

3.3.1 Techniques for Determining Stability Constants

Numerous methods have been employed to determine cation binding constants. Which of the available techniques is used depends on the reac-

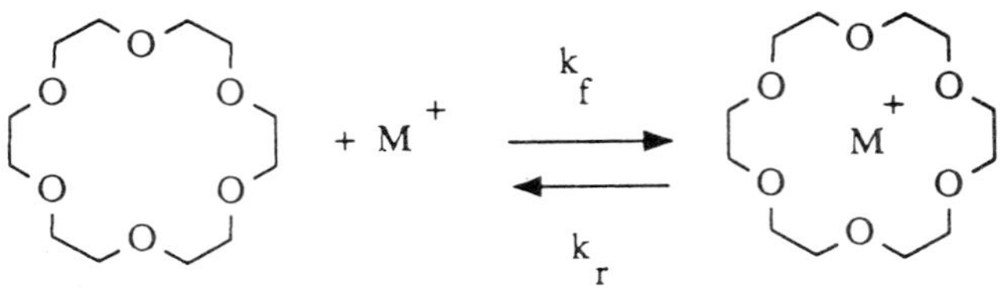

$$\text{18-crown-6} + M^+ \underset{k_r}{\overset{k_f}{\rightleftharpoons}} [\text{18-crown-6}\cdot M^+]$$

Figure 3.3 *Cation complexation equilibrium for 18-crown-6 and a metal cation (M^+)*

tion under study and also, to some extent, on the group undertaking the problem. It is often the case that a substantial expertise exists within a research group and the technique is recognized as having a potential application in a certain area. Thus it is with macrocycle complexation chemistry. A number of experts in techniques have brought to bear both their interest and expertise on macrocycle complexation problems. Four techniques have been used most often for this purpose.

Equilibrium stability constants may be determined in homogeneous solution using *conductance* methods. The experiment is usually conducted on a solution of an alkali metal salt in an appropriate solvent in a constant temperature bath. The resistance of this solution is determined in a cell designed for the purpose. Crown ether is added incrementally until about a five-fold excess of the macrocycle has been added. Equilibrium binding constants are then calculated using a least squares analysis.

Titration calorimetry has been used extensively to obtain equilibrium binding constants although its principal value has certainly been in the determination of free energies of reaction (ΔG values) and the enthalpic (ΔH) and entropic ($T\Delta S$) components thereof. Experimentally, the technique involves use of a calorimeter and the heat liberated during titration of the macrocycle by salt or *vice versa* is accurately and precisely determined. After correction for heat effects due to processes other than the reaction, a procedure to obtain an iterative fit of the thermochemical data is undertaken. Values are approximated for the cation binding constant and corresponding values of ΔH are calculated until the latter are constant for all points on the thermogram. This method is most accurate when the cation binding constant has a finite value less than 10 000.

Because calorimetry uses this iterative fit method, sometimes the 'best' values of K_S are not the chemically most reasonable ones. Even so, the availability of ΔH and ΔS values by this technique more than compensates for any such deficiency. Reed Izatt, the late Jim Christensen and their co-workers have determined numerous cation stability constants and have collected the values available in the literature in two important *Chemical Reviews* articles. The latter of the two appeared in 1985 and was comprehensive when it appeared. The reader is directed to that article for supplementary information.

Cation binding constants can be determined by *nuclear magnetic resonance* (NMR) techniques and this method has been widely and successfully applied. The method depends on chemical shift or linewidth differences between complexed and uncomplexed species and may be used when such differences can be detected either in the ligand or the cation. The direct observation of such metal ions as ^{7}Li and ^{23}Na enhances the versatility of this technique. The chemical shift differences are observed in this technique and then related mathematically to the equilibrium constant for the temperature and solvent pertaining to the observation.

Probably the most versatile method for determining cation binding constants is the method developed originally by Pedersen and Frensdorff which uses *ion selective electrodes*. This technique is especially important because it utilizes relatively inexpensive equipment. An ion selective electrode, a constant temperature bath, and a sensitive voltmeter are required and a modest setup can cost about as much as a rotary evaporator. This certainly places the technique within reach of most research groups.

Ionic activities are determined for Na^+ using a sodium-selective electrode and for K^+ and NH_4^+ using a monovalent ion electrode. Divalent calcium can be determined using a calcium selective electrode but these are not generally useful in organic solvents so measurements must be confined to water. A competitive method has been developed so equilibrium constants can be obtained in a diversity of solvents.

The method involves determining the difference in EMF for a solution of salt in the presence and absence of ligand. The potential difference is substituted into the Nernst equation followed by application of mass balance equations. The equilibrium constant thus obtained is applicable only for the solvent and temperature at which it is determined. By measuring K_S at several temperatures and then applying the van't Hoff relationship, thermodynamic data of reasonable precision may be obtained.

3.3.2 Variables Affecting Homogeneous Stability Constants

Homogeneous stability constants differ from extraction constants in the essential sense that both ligand and salt are present in the same medium at all times. Many different metal salts have been used in homogeneous cation binding studies but chloride is probably the most common anion. Many solvents have been used as well and the choice often depends on the measuring technique, the solubility of the ligand, and the value of the binding constant in that solvent. The latter is important since there is generally an inverse relationship between solvent polarity and K_S. Methanol is probably the most commonly used solvent, probably because it represents a middle ground in polarity and can accommodate both metal salts and organic ligands.

Binding constants for the family of unsubstituted crown ethers and several cryptands are shown in Table 3.2. All of the values were determined in anhydrous methanol. Note that the values are expressed as decadic logarithms ($\log_{10}$). Thus, 15-crown-5 binds K^+ ($\log K_S = 3.43$, $K_S = 2692$)

almost twice (1.6 times) as strongly as it binds Na^+ (log K_S = 3.23, K_S = 1698).

We have previously referred to the so-called 'hole size' concept which holds that when a cation's diameter and a crown ether's hole are the same size, the latter is selective for the former. The cavity of 15-crown-5, for example, is estimated to be 1.7–2.2 Å and the ionic diameter or Na^+ is 1.9 Å. Unfortunately this generalization does not really hold as shown below. 15-Crown-5 is actually selective, albeit by a small factor, for K^+ over Na^+.

It seems reasonable that the hole size idea would be most valid with the most rigid systems. A flexible ligand can accommodate a wider variety of cations than can a rigid one. In the case of flexible ligands, such factors as the cation's solvation enthalpy and ligand conformations become important. Potassium cation has the lowest solvation enthalpy in the group Na^+, K^+, Ca^{2+} and is selected by all of the crowns shown in Table 3.2. 18-Crown-6 probably binds K^+ more strongly than does 12-crown-4 in part because it has a larger number of donors groups. 18-Crown-6 is also relatively strain free and has a highly symmetrical (D_{3d}) binding conformation (Figure 3.4).

The hole size relationship is better represented by the cryptands which are relatively rigid species. Thus [2.2.1]-cryptand has a cavity size similar to Na^+ and selects that cation over K^+. Larger [2.2.2]-cryptand has a cavity size similar to the ionic radius for K^+ and selects this cation over Na^+ or Ca^{2+}, both of which are smaller.

The reaction of NaCl with 15-crown-5 has been studied in mixtures of methanol and water. The reaction is illustrated in the general sense in Scheme 3.2. As the amount of water in methanol increases, so does the solvent polarity. The more polar solvent can compete more effectively with the ligand for the cation so the binding constant is lowered. The stability constants ($\log_{10}K_S$) are as follows: water, 0.8; 20% methanol, 1.49; 40% methanol, 1.71; 60% methanol, 2.21; 80% methanol, 2.65; 90% methanol, 2.97; and anhydrous (100%) methanol, 3.24. A plot of log K_S *vs.* mole

Table 3.2 *Cation binding strengths for crown ethers and cryptands*

	Stability Constants, log K_s			
Ligand	Na^+	K^+	Ca^{2+}	NH_4^+
12-crown-4	1.7	1.3	—	—
15-crown-5	3.24	3.43	2.36	3.03
18-crown-6	4.35	6.08	3.90	4.14
21-crown-7	2.54	4.35	3.27	2.80
24-crown-8	2.35	3.53	2.63	2.66
[2.2.1]-cryptand	9.4	8.45	9.92	—
[2.2.2]-cryptand	8.0	10.6	8.14	—
[3.2.2]-cryptand	4.8	>7	4.7	—

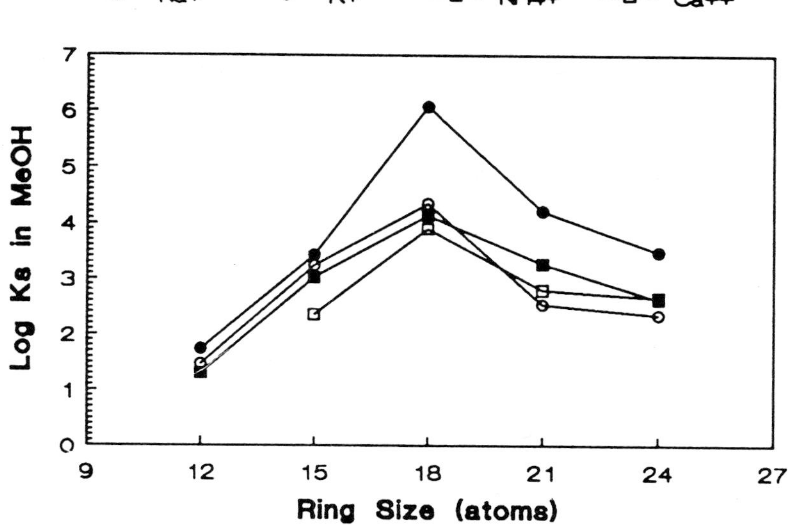

Figure 3.4 *Cation Binding by 3n-Crown-n Compounds*

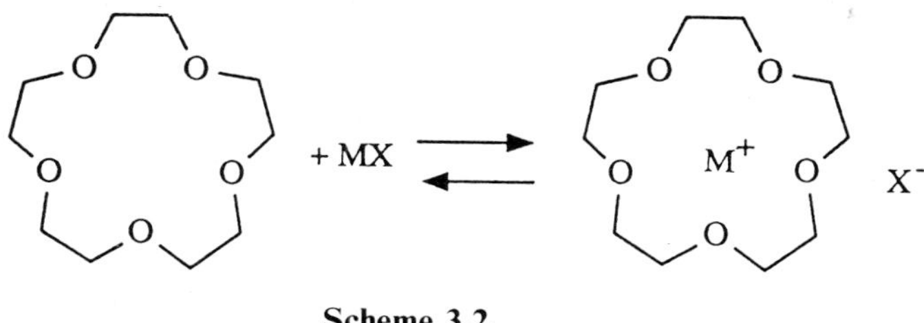

Scheme 3.2

fraction is nearly linear. The solvent polarity relationship suggested here does not hold for all ligands or all solvents but it is a useful generalization that is correct more often than not.

Michaux and Reisse used calorimetric methods to determine the thermodynamics of 18-crown-6 binding but independently determined K_S prior to fitting their data. As a result, their thermodynamic information differs slightly from the well-regarded values determined by Izatt and Christensen. The van't Hoff method described above was also used to determine thermodynamic components of binding and the values obtained by the three methods are compared in Table 3.3.

The conclusion is that each of these methods is useful. The three methods differ in some cases but are more similar than different. Using an experimentally-determined value of K_S has the advantage that the calculation needs to do less work to fit the thermochemical data. Either method is excellent but both suffer from the equipment requirement

Table 3.3 *Thermodynamic data for sodium ion complexation*[a]

Method	ΔH	$T\Delta S$	*log* K_S
15-Crown-5 binding Na^+			
Calorimetry/iterative fit	−4.99±0.03	−0.24	3.48
Calorimetry/iterative fit	−5.40±0.30	−0.90	3.30
Calorimetry/experimental K_S	−5.50±0.20	−1.23±0.24	3.14
van't Hoff method	−4.19±0.05	0.30±0.03	3.29
18-Crown-6 binding Na^+			
Calorimetry/iterative fit	−8.40±0.30	−0.24	4.36
Calorimetry/experimental K_S	−7.50±0.07	−1.55±0.11	4.37
van't Hoff method	−7.40±0.30	−1.50±0.09	4.34

[a]All values determined in anhydrous methanol at 25°C

inherent to calorimetry. It is in this latter context that the van't Hoff method is particularly attractive.

3.3.3 The Macrocyclic Effect

One of the earliest questions about cation complexation was 'what effect does ring formation have on binding strength?' The answer is that preorganization is significant indeed and is usually referred to as *the macrocyclic effect*. Only two examples need to be given to illustrate this phenomenon. The open-chained equivalent of 18-crown-6 is pentaethylene glycol dimethyl ether. Potassium (K^+) binding in anhydrous methanol solution ($\log_{10} K_S$) for the podand (**11**) is 2.3 (*i.e.* the equilibrium lies to the right and K is $10^{2.3}$ or 200). The corresponding macrocycle, 18-crown-6 (**10**) has a binding constant for K^+ in anhydrous methanol of ($\log_{10} K_S$) 6.08 ($10^{6.08}$ or 1 200 000). The macrocyclic effect on cation binding strength in this case is therefore about 6000-fold.

A similar study has been reported by Lehn. He showed that the K^+ binding constant ($\log K_S$) for [2.2.2]-cryptand (**13**) in 95% methanol (not the difference in solvent compared to 18-crown-6, above) was 9.75. The open-chained equivalent of this compound (structure **12**) has a K^+ binding constant in the same solvent of only 4.80. Indeed, Lehn's preparation of this interesting structure foreshadowed preparation of the lariat ethers. In this case, the macrocyclic effect is worth nearly 90 000-fold. Again, it is clear that preorganization is important for cation binding strength. The four compounds in question are shown in structures **10–13** along with their K^+ binding constants.

3.3.4 Enthalpy–entropy Compensation

Most chemists tend to think in 'enthalpic' terms. Such concepts as inductive stabilization and steric congestion are distinctly enthalpic. This

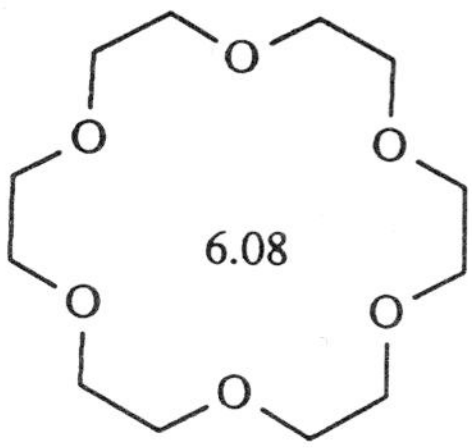

6.08

10

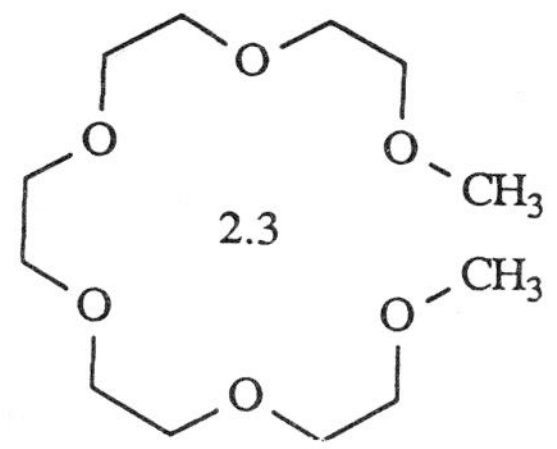

2.3

CH_3 CH_3

11

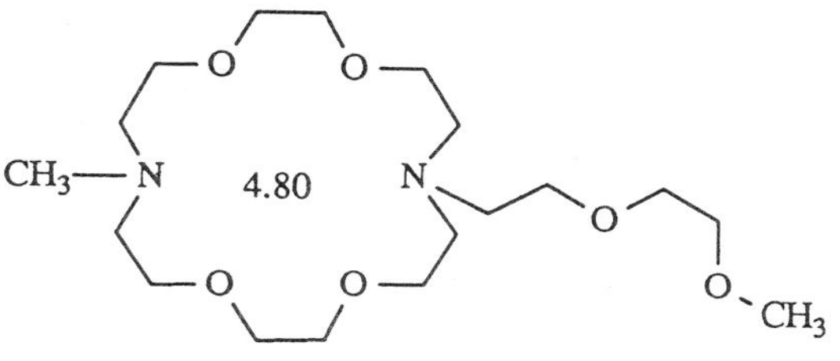

CH_3—N 4.80 N

CH_3

12

9.75

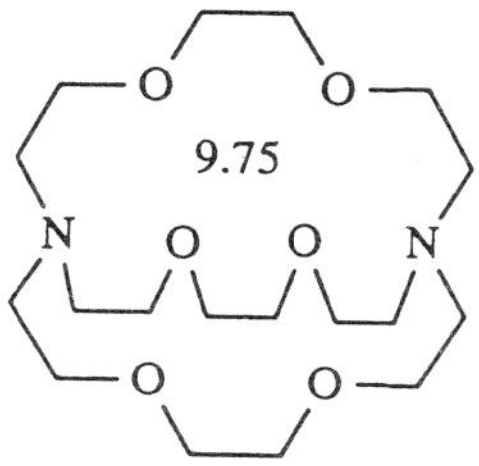

13

mode of thinking is prevalent despite the well known importance of solvent effects and organization. For most chemists, it is harder to think about the latter, however, and the enthalpic interpretation predominates in host–guest chemistry.

When enthalpy and entropy are both known, we may realistically consider the contribution of each to complex stability. Unfortunately, rarely is so much information available. For a number of compounds, both ΔH and ΔS have been reported and an interesting phenomenon has been observed. As the strength of the host–guest interaction increases (ΔH), solvent must also become more ordered (ΔS). Increasing ΔH is stabilizing but decreasing the randomness of solvent is not. Thus, cation binding constants may differ surprisingly little even though one expects a much stronger ligand–cation (host–guest, enthalpic) interaction in one case than in another. An example of this can be found in a comparison of several bibracchial lariat ethers based on the 4,13-diaza-18-crown-6 system (Scheme 3.3).

In Table 3.4, data for cation binding by a variety of structures is shown.

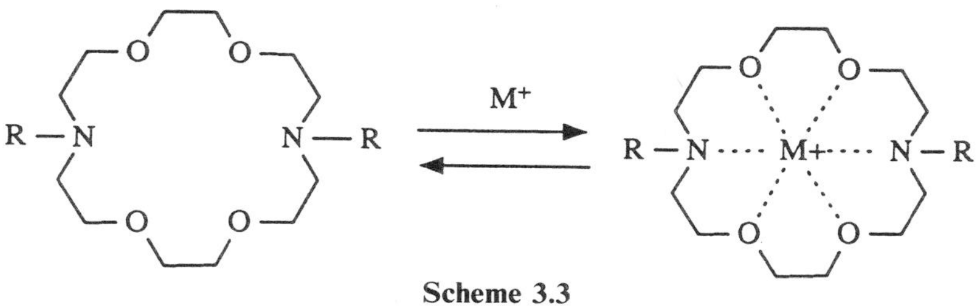

R—N N—R $\xrightarrow{M^+}$ R—N···· M^+···· N—R

Scheme 3.3

Table 3.4 *Binding constants and thermodynamic parameters for BiBLEs*

Compound or BiBLE Sidearm	*Cation*	*log K_S*	ΔH	$T\Delta S$
18-Crown-6	Na^+	4.35	−7.40±0.11	−1.50±0.09
	K^+	6.08	−11.3±0.02	−3.03±0.04
R = CH_2–C≡CH	Na^+	3.61	−4.97±0.04	−0.05±0.12
	K^+	4.99	−10.10±0.15	−4.21±0.09
R = CH_2–C≡N	Na^+	2.69	−4.87±0.08	−1.20±0.10
	K^+	3.91	−9.54±0.11	−3.29±0.17

A comparison of Na^+ and K^+ binding for each ligand reveals that in each case, ΔH is more negative for the latter. On the other hand, so is $T\Delta S$ so the cation binding constants do not differ by as much as might otherwise be the case. This is the essence of *enthalpy–entropy compensation*.

A comparison of the two BiBLEs having isosteric CH_2–C≡CH and CH_2–C≡N sidearms is also quite instructive. The Na^+ and K^+ binding constants for the propargyl derivative are both higher than for the nitrile. One is tempted by these data to assert that there is a π-interaction between the triple bond and the ring bound cation and that it is more effective for the C≡C bond than for the C≡N bond. This may be so, but it is not borne out by the data. Indeed, the enthalpic components of the interaction in both cases are very similar. The difference in the binding constant arises almost entirely from the entropic term.

An effort has been made by several groups to understand the thermodynamics of host–guest interactions, but the range of compounds involved in these studies has not proved to be too broad. This is unfortunate since ligand design could benefit from a more detailed understanding of solvent's role in stabilizing the free ligand and its various complexes.

3.3.5 Brief Survey of Cation Complexation Constants

An extensive collection of cation binding data have been accumulated by Izatt *et al.* (see above and references). It is well beyond the scope of the current monograph to duplicate that information but it is useful to collect a few data here. By so doing, we may gain some insight into the factors that affect cation binding strengths and selectivities (Table 3.5).

Although it is difficult to generalize from such a limited data set, some trends are clear. Placement of a benzo group in a 15-crown-5 residue decreases Na^+ binding. This is probably because the sp^2 hybridized oxygen atoms are less basic than are the sp^3 hybridized oxygen atoms in the parent and the sp^2-hybridized electron pairs are perpendicular to the aromatic ring. Increased hydrophobicity may be playing a role as well because cyclohexano-15-crown-5 binds Na^+ cation more strongly than

Table 3.5 *Typical Homogeneous Cation Binding Constants*[a]

Ligand	*Complexation Constant*			
	Na^+	K^+	Ca^{2+}	NH_4^+
12-crown-4	1.7	1.3	—	—
15-crown-5	3.24	3.43	2.36	3.03
benzo-15-crown-5	3.1[b]	3.0[b]	—	—
cyclohexano-15-crown-5	3.71	3.58	—	—
aza-15-crown-5	2.06	2.72	—	3.05
N-*n*-butylaza-15-crown-5	3.22	2.99	2.8	—
N-($CH_3OCH_2CH_2$)aza-15-crown-5	4.33	4.20	3.78	3.15
18-crown-6	4.35	6.08	3.90	4.14
benzo-18-crown-6	4.3[b]	5.3[b]	3.50	—
cyclohexano-18-crown-6	4.09	5.89	—	—
Z-*syn*-*Z*-dicyclohexano-18-crown-6	4.08	6.01	—	—
E-*anti*-*E*-dicyclohexano-18-crown-6	2.52	3.26	—	—
E-*syn*-*E*-dicyclohexano-18-crown-6	2.99	4.14	—	—
dibenzo-18-crown-6	4.4[b]	5.0	—	—
aza-18-crown-6	2.77	4.18	—	—
N-*n*-octylaza-18-crown-6	3.51	4.65	—	—
N-($CH_3OCH_2CH_2$)aza-18-crown-6	4.58	5.67	4.34	4.12
4,13-diaza-18-crown-6	1.5	1.8	—	—
4,13-(*n*-propyl)$_2$diaza-18-crown-6	2.86	3.77	—	—
4,13-($CH_3OCH_2CH_2$)$_2$diaza-18-crown-6	4.75	5.46	4.48	—

[a]In anhydrous methanol at 25°C. [b]Average of two or more published values

either. Stereochemistry can play an important role. The difference among the dicyclohexano-18-crown-6 isomers is striking.

The crystal structures of di- and triaza-18-crown-6 show that the >N–H bond is turned inward. This presumably helps prevent any cation from occupying this space. The >N–H group must also be extensively hydrogen bonded in methanol solution and this is reflected in the diminished stability constant values. Nitrogen, of course, exhibits a reduced affinity for alkali metal cations as expected compared to the all-oxygen analogs. It should be noted that few cation affinity data are reported for diaza-18-crown-6 with alkali metal or alkaline earth cations but data are reported for the following transition elements: Co, Ni, Cu, Ag, Zn, Cd, Hg, and Pb. A general rule is that more oxygen atoms favor group IA and IIA elements and a greater number of nitrogen atoms favors transition and coinage metals. Indeed, tetraaza-14-crown-4, cyclam, is one of the most studied of all known ligands and is the subject of numerous reports even today.

3.4 Binding Dynamics

One reason many regard homogeneous cation binding constants as generally more useful than extraction constants is that the thermodynamics and binding dynamics are obviously related. The multiple equilibria inherent to extraction constants makes such assessments more difficult.

Since the relationship of binding strength and rate is

$$K_S = k_{complex}/k_{decomplex}$$

only two of the three variables need to be determined. Several methods (detailed previously) exist for the determination of K_S, so the rate for only one component of this equilibrium needs to be determined. This is important since the binding and release rates range over many powers of ten and are difficult to measure in the fastest cases.

The crown ethers are generally much more flexible than are the cryptands since the former are monocyclic and the latter are tied together by a third molecular strand. As a result, complexation and decomplexation rates are typically faster for crown ethers than for cryptands. Of course, these rates are influenced by such factors as temperature, solvent, and cation. The binding rates must be determined by appropriate fast kinetic techniques such as ultrasound and in this connection Eyring, Petrucci, Liesegang, and others have been important contributors. NMR has also proved useful for understanding complexation and dynamic properties although fast rates are more difficult to obtain by this technique. Popov, Chan, Echegoyen, and Kaifer and their co-workers have been particularly active in this effort. Some representative rate data are shown in Table 3.6.

Generalizations from these data are difficult to make but one concludes that for the most part, crown ether complexation is more facile (faster rates) than cryptand complexation. This is expected based on the relative rigidities of these two ligand systems. It also appears that binding

Table 3.6 *Complexation and decomplexation rates for crowns and cryptands*

Ligand	*Solvent*	*Cation*	$k_{complex}$ $M^{-1}s^{-1}$	$k_{decomplex}$ $M^{-1}s^{-1}$
15-crown-5	H_2O	Na^+	2.6×10^8	5.1×10^7
15-crown-5	H_2O	K^+	4.3×10^8	5.1×10^7
18-crown-6	H_2O	Na^+	2.2×10^8	3.4×10^7
18-crown-6	H_2O	K^+	$4.3–6.3\times10^{10}$	$3.7–12\times10^6$
18-crown-6	MeOH	K^+	6.8×10^5	36
[2.1.1]-cryptand	H_2O	Li^+	$1–8\times10^3$	$5–25\times10^{-3}$
[2.1.1]-cryptand	H_2O	Na^+	9×10^4	1.4×10^2
[2.2.1]-cryptand	H_2O	Na^+	$3.6–6\times10^6$	14.5–18
[2.2.1]-cryptand	H_2O	Ca^{2+}	$6–16\times10^3$	$6.6–22\times10^{-3}$
[2.2.1]-cryptand	H_2O	K^+	3×10^7	2×10^3
[2.2.2]-cryptand	H_2O	Na^+	2×10^5	27
[2.2.2]-cryptand	H_2O	K^+	$2–8\times10^6$	9–42
[2.2.2]-cryptand	MeOH	K^+	4.7×10^8	1.8×10^{-2}

rates (k_C) for the same ligand with different cations are slower for the more charge dense ions. Again, this is expected if the charge-dense ions are bound more strongly by solvent than the more charge-diffuse ions.

3.5 Cation Transport

Transport of cations across a phase boundary has been one of the most studied applications in the crown ether and cryptand area. The reaction is often conducted in a 'U-tube' because of its shape. This simple device is sometimes called a 'Pressman' cell after Berton Pressman (Figure 3.5). Its use is fairly straight-forward and the principles are, to some extent, borrowed from the 'extraction constant' notions.

In a typical experiment, the lower half of the U is filled with chloroform or other organic solvent having density >1 g ml^{-1}. Water is placed in both arms, one of which will serve as source phase and the other of which will serve as the receiving phase. If the source phase contains sodium, it may be transported by a crown ether in the chloroform phase. Naturally, when the cation is transported, an anion must accompany it. The anion is often chosen to be UV-active or fluorescent so that aliquots of either the source or receiving phase can be examined at intervals to determine the rate of appearance or disappearance of salt. Of course, such techniques as atomic absorption may be used to observe the change in cation concentration directly. The experimental apparatus is illustrated schematically in Figure 3.5.

An alternative design that is clever and equally useful involves a beaker with a glass tube inserted. The bottom half of the beaker is filled with the bulk organic membrane and then a glass tube is inserted. The interior of the tube above the organic solvent is filled with the source or receiving

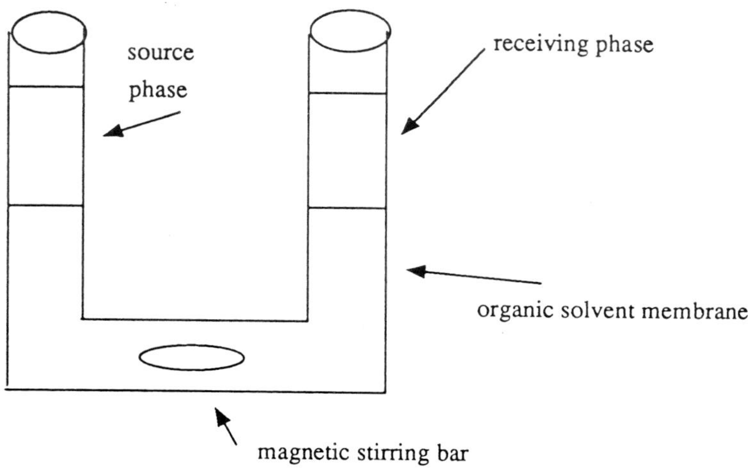

Figure 3.5 *A 'Pressman' cell*

solution and the outer cylinder is filled with the complementary phase. This device is probably easier to construct and the membrane surface area is easy to adjust simply by changing the diameter of the inner tube or vessel (Figure 3.6).

Many transport studies have been conducted as they are a useful and convenient way to assess ionophore activity within a series. As with extraction constants, comparisons with other systems must be made under identical experimental conditions for the comparison(s) to be valid. The most useful information that can be obtained from this type of experiment is probably the actual rate at which cations are transported from the source to the receiving phase. Unfortunately, this is done only in a minority of cases. It is more common to conduct the transport experiment for a specified length of time and report only the percent of available cation observed in the receiving phase. While such information is useful, a rate constant determined under standard conditions would be better.

Other variables that are of concern in transport processes are the ratios of volumes (membrane:source and/or receiving phase), concentration of ionophore, contact area between aqueous and organic phases, and stirring rate. The stirring rate is especially important since it should be fast enough to ensure efficient mass transport without emulsifying the separate phases. If the membrane and aqueous phase(s) become emulsified, the contact area may change drastically.

In principle, passive transport can occur only to the extent of 50% since the driving force for transport is lost in the absence of a concentration gradient. For this reason, transport is often assessed only during its initial

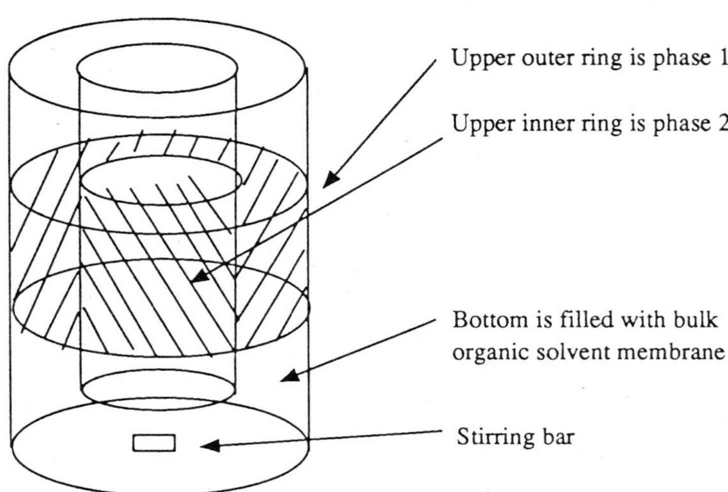

Figure 3.6 *Alternative apparatus for cation transport measurement*

stages. Transport of a cation can also be driven by a pH gradient and this approach has been used with both crowns and cryptands. A schematic representation of cation transport is shown in Figure 3.7.

3.6 Complexation of Organic Cations

Pedersen recognized from the very beginning that species other than metallic cations might be complexed. He demonstrated the complexation of ammonium ions in his earliest papers and even demonstrated the inclusion of thiourea. The complexation of species such as diazonium ions ($-N_2^+$), acylium ($-C{=}O^+$), hydrazinium, hydronium, and many others have been demonstrated. In these cases, stabilization involves a dipolar interaction with a cationic species which may involve direct hydrogen bonding. Many other complexes have also been identified such as the acetonitrile and nitromethane complexes but these appear to be molecular inclusion complexes rather than complexes that can be detected in solution.

3.6.1 Ammonium Cations

Alternating oxygen atoms of an 18-membered crown ether ring are appropriately spaced to hydrogen bond with each of three N-H bonds of an ammonium ion (II).

The tetrahedral nitrogen forms a tripod of bonds to the hydrogen bond acceptor groups. The 18-crown-6 ring exhibits D_{3d} symmetry in this complex. The nitrogen atom resides above the mean plane of the oxygen atoms and the fourth bond, shown in **14**, to an arbitrary residue, R is perpendicular to the plane of the ring. Such complexes have proved quite stable in many cases and the interaction has been used to permit chiral recognition and resolution in certain cases.

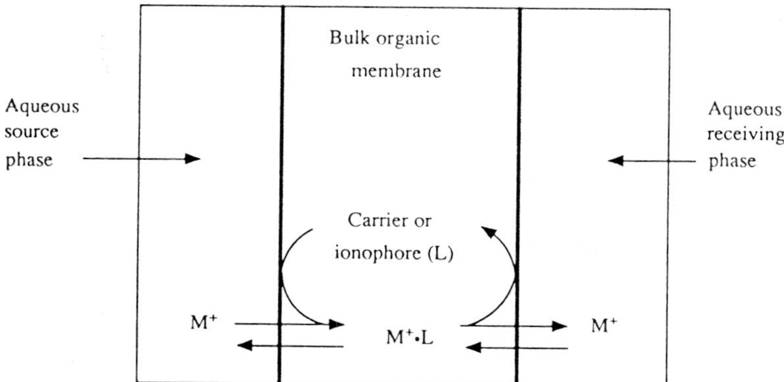

Figure 3.7 *A schematic representation of cation transport*

14

CPK atomic model of 18-crown-6 complexing the tetrahedral ammonium cation, NH_4^+

The N–H–O interaction has been used extensively in complexation chemistry (as detailed previously) and some interesting variations on the ammonium complex have been prepared. Cram and co-workers, for example, have demonstrated 1:1 complexation between guanidinium ion and 27-crown-9 (**15**).

In this complex (**15**), six hydrogen bonds are believed to stabilize the system. Three of the oxygen atoms serve a role as spacers but, if they were replaced by methylene groups, the entire complex might be destabilized by a change in the system's overall conformational requirement. To our knowledge, no crystal structure of this complex has appeared but complexes of similarly stabilized systems have been studied.

The hydrogen atoms of urea are somewhat less acidic than those of an ammonium cation and the binding is consequently poorer. Reinhoudt and co-workers have overcome this difficulty by incorporating an acidic function in position to hydrogen bond (partially protonate) the urea carbonyl group as in structure **17**. This interaction forms an anchor point

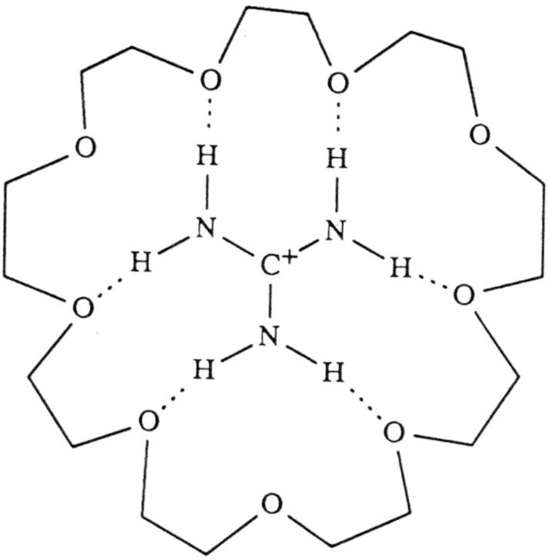

15

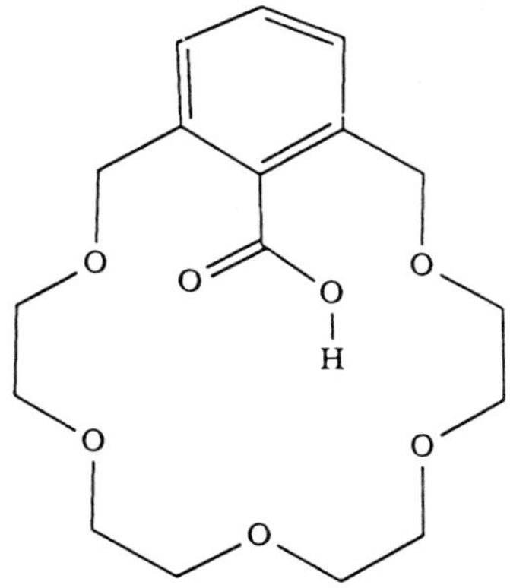

16

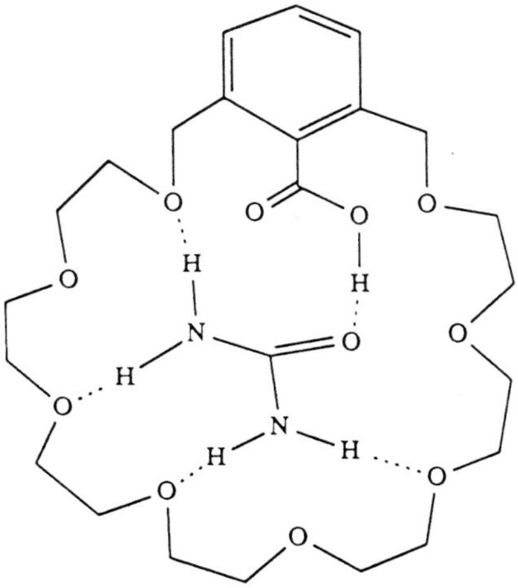

17

within the macrocyclic structure while generally acidifying the >N–H bond. Of course, the size of the ring is critical. The ring shown in structure **16** is inappropriate and could not accommodate urea even though a carboxyl group is present.

Another clever example of ammonium cation binding was reported by Stoddart, Colquhoun, and their co-workers at Sheffield and Imperial Chemical Industries. In this case, a transition metal cation is bound to ammonia (the inorganic ammine ligand) (**18**). The transition metal acidifies the N–H bonds and these are readily complexed by a crown ether. This type of complexation has been dubbed 'second sphere solvation.'

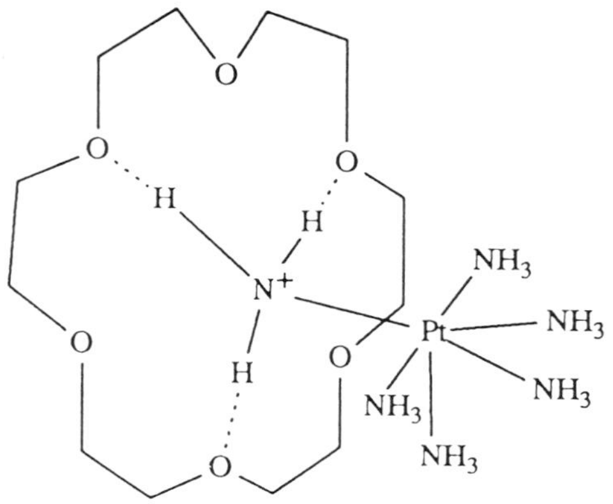

18

3.6.2 Chiral Recognition

As early as the late 1960s, Cram envisioned what he referred to as an 'amino acid resolving machine'. The plan was to use the tripod of hydrogen bonds that form between an ammonium salt and a crown ether. The ammonium salt would be the protonated form of an amino acid, H_3N^+–CHR–COOR′. The crown would be so modified as to contain a 'chiral barrier', that is an asymmetric steric environment that would favor one amino acid enantiomer over the other. Cram hoped that only the favored enantiomer in a racemate would complex the macrocycle and dissolve. Resolution would thus be achieved by filtration. Although this precise vision was never realized, the general principle has been convincingly demonstrated.

Cram chose for his chiral barrier the binaphthyl residue, a unit that also intrigued Nobel Laureates Vladimir Prelog (Zurich) and Jean-Marie Lehn (Strasbourg). He used two binaphthyl residues in optically active form across from each other in a six-oxygen containing ring.

The hydrogen atoms in the 8-position of the naphthalene rings (the *peri* hydrogens) cannot occupy the same space so the naphthalene rings cannot pass each other at normal temperatures and pH values. When complexation occurs, the chiral residue will be most stabilized when the large (L) residue is in the largest cavity, and the smallest residue (S, usually hydrogen) is directly against the naphthalene ring system. Indeed, it is precisely this that Cram and co-workers observed. Using an optically pure (*R,R* or *S,S*) macrocycle, complexation of racemic salts such as α-phenethylammonium was observed by NMR. The macrocycle was soluble in $CDCl_3$ and one equivalent of the salt partitioned from D_2O into it. Since one diastereoisomeric complex was more stable (because of fewer unfavorable steric interactions) than the other, a predominance of it was observed. In the case described here (**19**), two pairs of methyl proton reso-

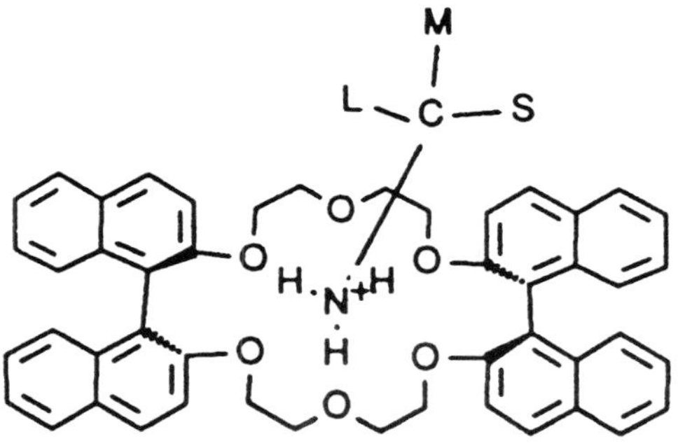

19

nances were integrated to assess the amounts of complex present. In practice, *R,R* macrocycle favored the *S*-isomer of phenethylammonium PF_6^-.

Ultimately, the macrocycles were bound to silica support surfaces and various racemic salts were chromatographed over the system. Since the chromatographic process increased the number of complexation interactions (theoretical plates), a complete resolution could be effected, albeit on a small scale.

3.6.3 Arenediazonium Cations

Cram also recognized that cylindrical cations such as diazonium ion might be complexed by the circular and polar cavity of a crown ether. Arenediazonium cations were selected for study and the anticipated interaction confirmed by a variety of techniques. Arenediazonium tetrafluoroborate (BF_4^-) or hexafluorophosphate (PF_6^-) salts are much more stable than the corresponding $Ar{-}N{\equiv}N^+$ halide, bisulfate, or nitrate salts. They are also essentially insoluble in chloroform and dichloromethane. Addition of 18-crown-6 to chloroform in contact with solid $Ar{-}N_2^+BF_4^-$ stoichiometrically dissolved the salt (**20**). The spectrum of the (complexed) salt could be observed in the NMR. 4-Methylbenzenediazonium tetrafluoroborate was studied because the proton NMR of its aromatic region is a sharp AB quartet making it readily distinguishable. When the methyl group was moved to the *ortho*-position, it was expected to interfere with complexation about the cylindrical cation and, indeed, the salt remained insoluble in chloroform, even in the presence of 18-crown-6.

Another indication of molecular complexation was the fact that partially insoluble salts completely dissolved when the temperature of the solution was *lowered*. This indicates that the large entropic component ($T\Delta S$) of complexation became less unfavorable as the temperature was lowered.

18-Crown-6 was chosen for these studies because CPK molecular models indicated a nearly perfect fit of macroring cavity and diazonio group upon complexation. Models also suggested that 15-crown-5 would be too

O O O $N^+{\equiv}N$ PF_6^- O O O

20

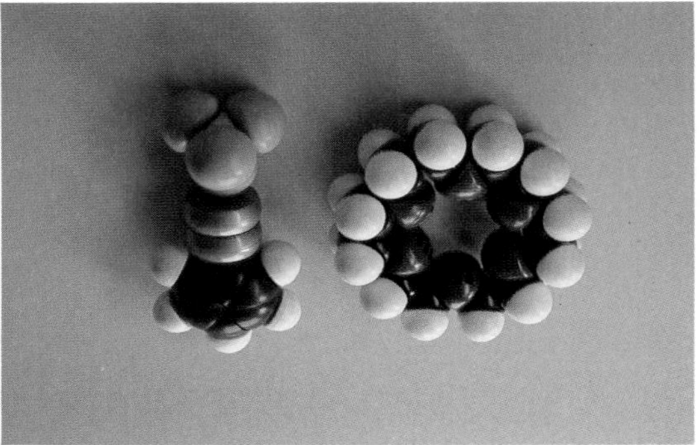

(a) CPK atomic model of benzenediazonium tetrafluoroborate (BF_4^-) next to 18-crown-6

(b) CPK atomic model of the complex between 4-methylbenzenediazonium cation and 21-crown-7

small to complex this residue and no evidence either of complexation or solubilization of these systems was ever observed. In extensive kinetic studies, Bartsch and co-workers showed that complexation of arenediazonium cations was most effective when 21-crown-7 was used as ligand. Izatt, Christensen, and their co-workers showed that the thermodynamic components of complexation for 4-methylbenzene-diazonium tetrafluoroborate complexed by 18-crown-6 were as follows: $\Delta H = 2.26$ kcal mol^{-1} and $T\Delta S = -6.95$ cal deg^{-1} mol^{-1}.

Several attempts were made to capture the macroring-bound arene-diazonium cation as a rotaxane. Reagents such as *N*, *N*-dimethylaniline and diphenylzinc were used in the hope that the azobenzene derivative thus formed would be threaded through the macrocycle. The hoped-for rotaxane (shown in structure **21**) was never obtained even though the coupling products did form. This suggested that the decomplexation rate was too fast even at quite low temperatures for this approach to be practical.

Evidence for complexation of acylium cations, R–C$^+$=O, was also obtained but this cation is so electrophilic that complexes were difficult to characterize.

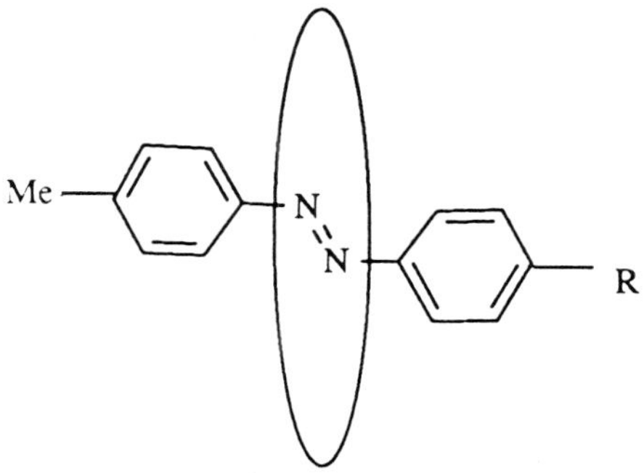

Me N N R

21

3.6.4 Hydronium Cation

The hydronium cation, H_3O^+ is similar in size and shape to the ammonium cation although it may be somewhat more planar. The three hydrogen bonds radiate at an angle of ~120° from each other. Izatt, Haymore, and Christensen described the first hydronium ion complex in 1972: the hydronium perchlorate complex of dicyclohexano-18-crown-6. The presumed structure is shown on the left in Figure 3.8. The complex was characterized by infrared spectroscopy but no *X*-ray crystal structure was obtained at that time.

In 1977, we isolated a complex that we believed for a time was a hydronium complex and might be. It is drawn in Figure 3.8 as a hydrated ammonium complex but it is frankly difficult to distinguish an N^+–H · · · O linkage from an N · · · H–O^+ bond. Most hydrogen bonds are double well potentials, that is, they may be close to either electronegative element. If the structure is not the first crystal structure of a hydronium cation complex, it is the first complex of a neutral water molecule.

Since these two early complexes were reported, numerous other complexes of organic species have been discovered. Some of these are mentioned in Section 3.7.

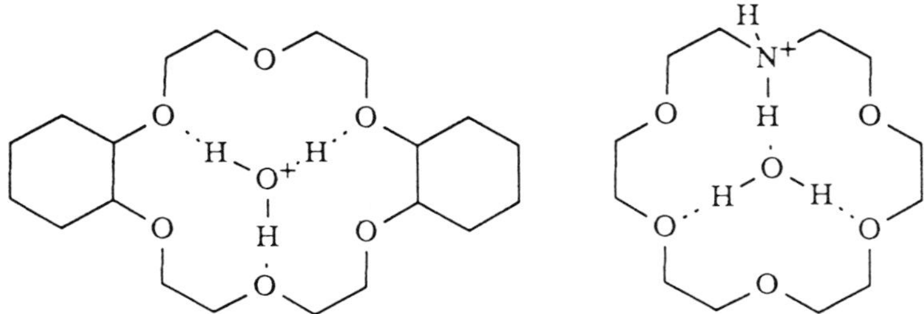

Figure 3.8 *Structures of hydronium perchlorate complexed by dicyclohexano-18-crown-6 (left) and aza-18-crown-6 complexing the elements of H_3OCl (right)*

CPK atomic model of 18-crown-6 complexing a water molecule

3.6.5 Co-operative Structures

Reinhoudt has complexed urea by using a clever variation on simple ammonium cation binding. The salen-crown system shown in structure **22** complexes urea by using a combination of donor interactions and Lewis acid chemistry. Formally, a simple crown ether is opened and the salen unit is inserted. The phenols are readily dissociated and the system is appropriately sized to accommodate the large cation uranium dioxide. The uranyl cation is capable of expanding its coordination number and readily accommodates the oxygen atom from urea (**23**). This acidifies the urea molecule, enhancing binding strength in the ligand.

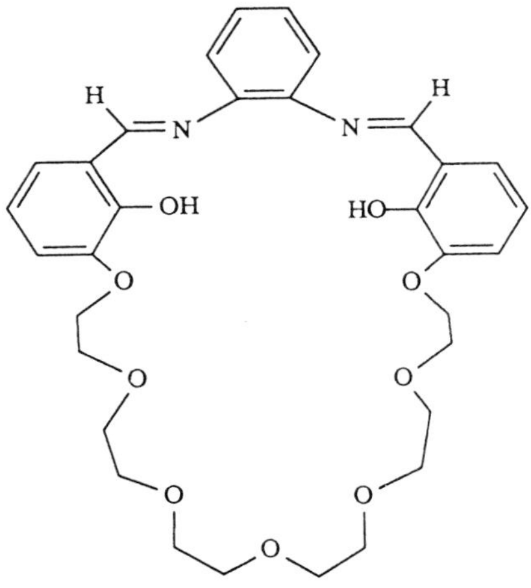

22 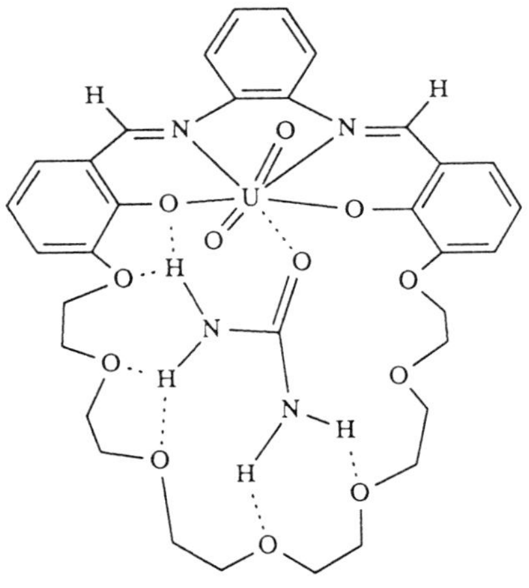 **23**

Another example of organic cation complexation that involves co-operativity is found in the 'second sphere complex' studies first reported by Stoddart and co-workers. In this case, the ammine ligand, NH_3, forms a complex to the crown and also is a ligand for a cation such as platinum. The Pt atom is thus held in proximity to the crown by a tripod of hydrogen bonds. The N-H bonds are acidified by the formal positive charge on nitrogen that results from coordination as follows: $H_3N \rightarrow Pt$. Note that in the ammine→Pt complex, the hydrogen bonds are not illustrated. Coordination of water is also possible by this phenomenon. Three illustrative structures are shown in Scheme 3.4.

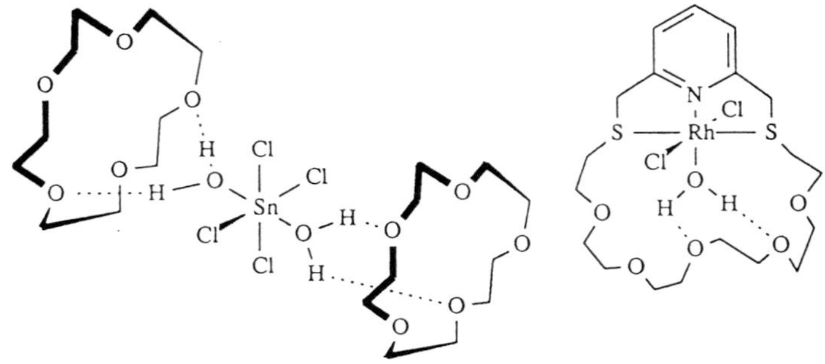

Scheme 3.4a

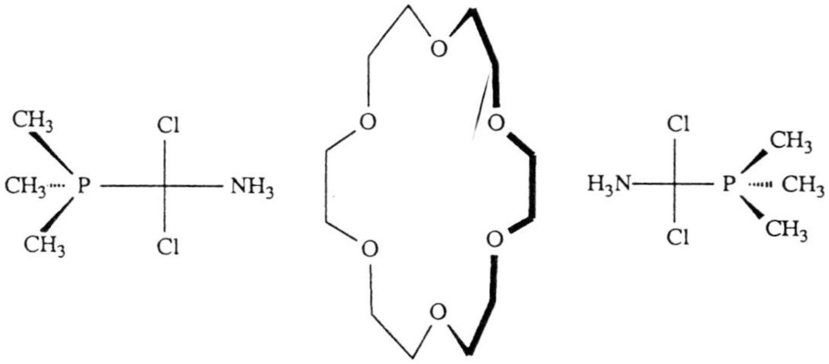

Scheme 3.4b

3.6.6 A Polymeric Ionophore

Novak and Grubbs have reported an interesting olefin metathesis reaction that converts the 7-oxanorbornene system shown in Scheme 3.5, into a polymeric ionophore. While not strictly a crown ether, it can be described as a coiled podand and the poly(donor group) properties are obvious. Indeed, transport experiments were conducted and three alkali metal cations showed an order $K^+ > Na^+ = Li^+$.

3.7 Molecular Complexation

Most of the complexes discussed above are stoichiometric complexes of metallic or organic cations in which the positively charged species interacts by a pole–dipole mechanism or by hydrogen bonding. Another general class of complexation exists which might be most properly termed *molecular inclusion*. In such 'complexes' the macrocycles pack in the lattice with voids filled by smaller molecular species.

The first such example is the thiourea inclusion complex observed by Pedersen in 1971 although its exact structure was not reported. The first complex that proved to be of practical interest is the so-called acetonitrile complex. When partially purified 18-crown-6 is stirred with acetonitrile, a

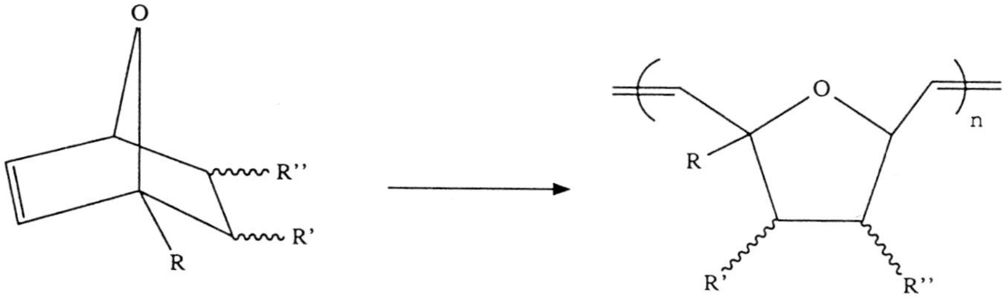

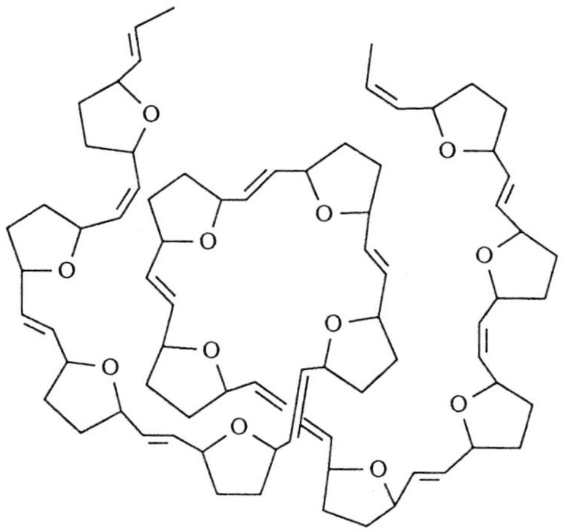

Scheme 3.5

solid forms. The solid has varying stoichiometries but appears to be an inclusion complex (solvate) in all cases. It can be separated and the acetonitrile removed to reveal highly pure 18-crown-6. The complex was first reported by Cram, Gokel, Liotta, and co-workers in 1974 but the crystal structure was not reported until 1988. Coincidentally, Garrell and associates working with Fronczek, who has studied the structures of numerous complexes, and Rogers and co-workers independently reported the structure of the acetonitrile complex in the same issue of the *Journal of Inclusion Phenomena*.

The nitromethane complex is similar and was discovered in the laboratories of Royal Dutch Shell by Reinhoudt and co-workers. By 1978, reports of dimethylthallium, benzenesulfonamide, and acetone complexes had all appeared. Charge transfer complexes of TCNE and 1,3,5-trinitrobenzene had also been reported.

An especially interesting complex was reported by Cram and Goldberg. It is the complex of crown with dimethyl acetylenedicarboxylate, $CH_3OCO{-}C{\equiv}C{-}COOCH_3$. In the solid state structure, the methyl group hydrogens are within hydrogen bonding distance of the oxygen donor atoms. Thus a complex that is apparently an inclusion complex is a candidate for a stoichiometric complex although, to our knowledge, no definitive demonstration of 1:1 complexation between a crown ether and a methyl group has ever been reported.

Weber and others have made a special study of the neutral complexes. They have synthesized a broad range of macrocycles in an effort to understand which variations in the individual ligands are reflected in the unit cell of crystals. It is hoped that these systematic studies will lead to an understanding of how individual ligands will affect cavities within a macroscopic solid. Design of such cavities would be of obvious interest in the custom synthesis of materials for a variety of applications.

Most of these complexes are fortuitous in the sense that precise intermolecular interactions were not anticipated and then designed into a specific structure. Increasingly, such efforts are being undertaken, however. Two such macrocyclic systems are illustrated in Schemes 3.6 and 3.7. In the first, Cram and co-workers built a spherand having one urea and two amide donor groups. As envisioned, the primary hydrogen bonding interaction occurs between the urea oxygen and the >N–H bond of imidazole.

Vögtle and co-workers have prepared a number of large ring macrotricyclic aromatic systems. These might be called cyclophanes or aromatic cryptands as they have structural features related to each. Indeed, this illustrates the merger of these sub-disciplines. In a recent report, intramolecular donor groups have been incorporated into the aromatic cyclophanes increasing the similarity to cryptands. When six nitrogen atoms are present as shown in Scheme 3.7, 1,3,5-trihydroxybenzene (phlorglucinol) is bound in dichloromethane solution. As in the imidazole complex in Scheme 3.6, the structure is illustrated as the binding might occur.

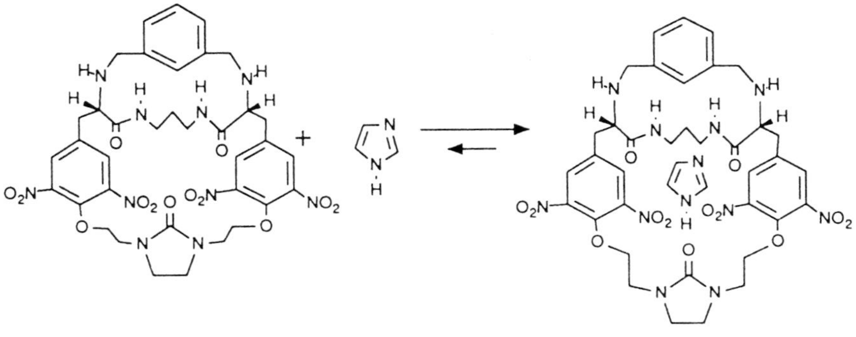

Scheme 3.6

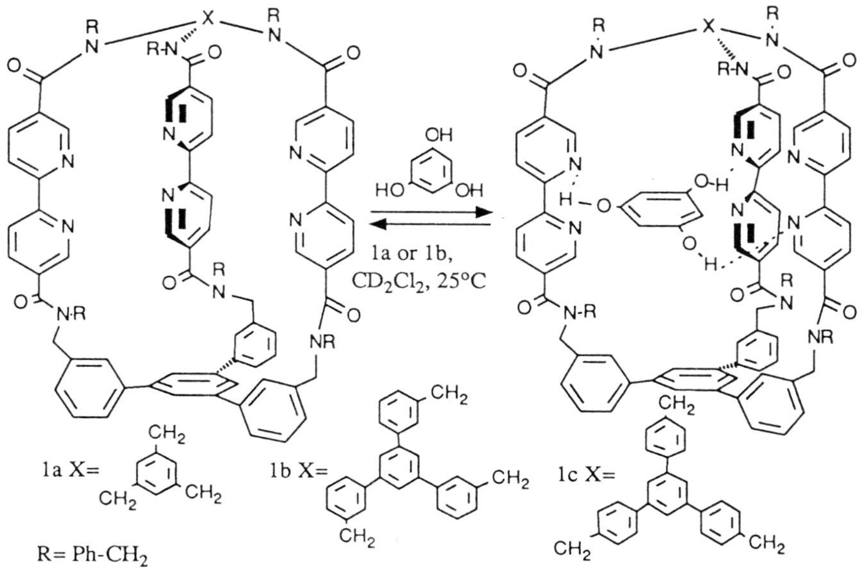

1a or 1b,
CD_2Cl_2, 25°C

1a X=

1b X=

1c X=

R= Ph-CH_2

Scheme 3.7

3.8 Anions

It seems a logical step to progress from the complexation of cations and neutral species to the entrapment of anions. Simple macrocycles are not well adapted to this task, however, as their heteroatoms are donor rather than acceptor groups. Some macrocycles containing main group elements such as tin have been prepared but such structures are beyond the scope of the present discussion. It is the cryptands rather than the crown ethers that have proved of interest with respect to anions.

One of the most important early developments was the effort made by Dye (Michigan State University) to completely encapsulate Na^+ and thus permit Na^- to exist. Dye used [2.2.2]-cryptand for this purpose since it, unlike most crown ethers, is an encapsulating ligand. The Na^- ion was reported in 1974, a cryptated Na^+ ion surrounded in the crystal lattice by six Na^- ions. The crystals are a beautiful golden yellow which must be due to the anion since the Na^+[2.2.2] complex is colorless. The stoichiometry of the complex is illustrated in Figure 3.9.

In subsequent work, Dye and his collaborators have extended this remarkable family of molecules to include the first example of an electride: a free electron residing in a void channel within the crystal.

Lehn and co-workers trapped a number of anions within cryptands by protonation. The complex of azide anion reported in 1978 is illustrated in the Figure 3.10.

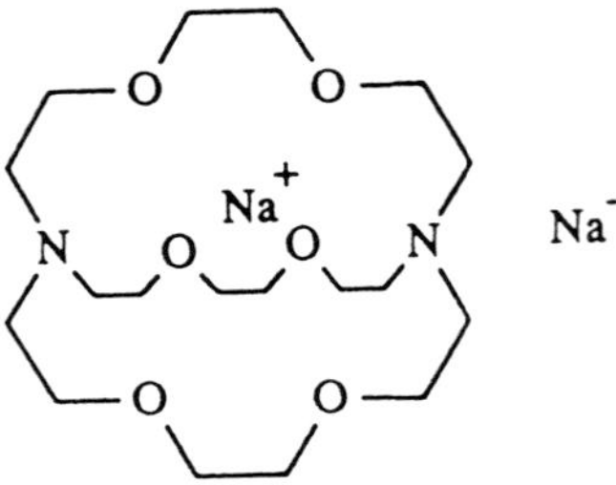

Figure 3.9 *The sodide ion (Na^-) with [2.2.2]·Na^+ as countercation*

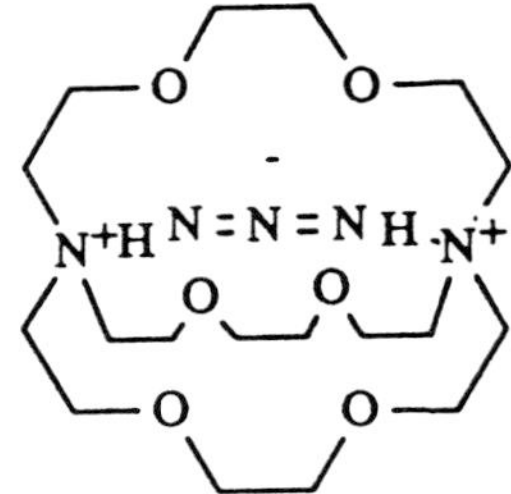

Figure 3.10 *Diprotonated [2.2.2]-cryptand complexing linear azide anion*

Chloride and other anions have been encapsulated using the same principle. One of the cleverest extensions of the idea involves an elaboration of the cryptand itself and was reported by Martell and co-workers. The extended cryptands are based on *tris*(aminoethyl)amine and are given the trivial name tren. When these complexes are protonated, they encapsulate copper ions which, in turn, trap molecular oxygen. The structure is shown in Figure 3.11.

One of a few examples are shown here and relatively few examples of anion complexes yet exist. This is a major area of research, however, and it is likely to see significant development in the foreseeable future.

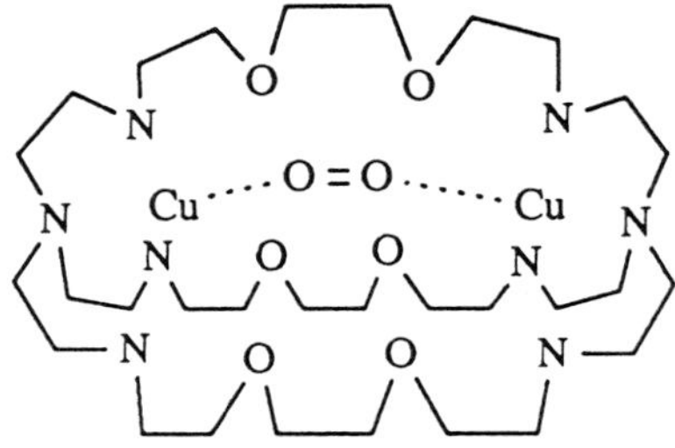

Figure 3.11 *Bistren complex of copper which, in turn, complexes dioxygen*

Hosseini, Blacker, and Lehn have reported a remarkable structure capable of complexing a nucleotide. The purpose of the study was to develop a receptor having more than one binding site (polytopic receptor) that might be able to bind ATP and catalyze its hydrolysis. The hexaaza (bistren) macrocyclic system shown in **24** can be hexaprotonated. The

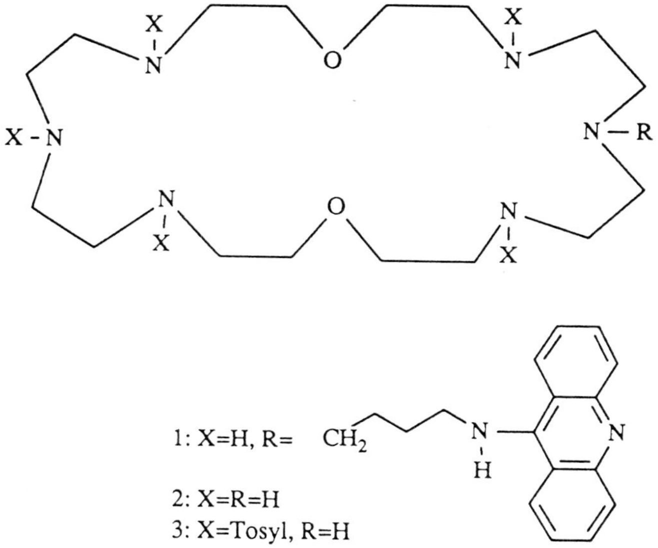

1: X=H, R= CH_2…

2: X=R=H

3: X=Tosyl, R=H

24

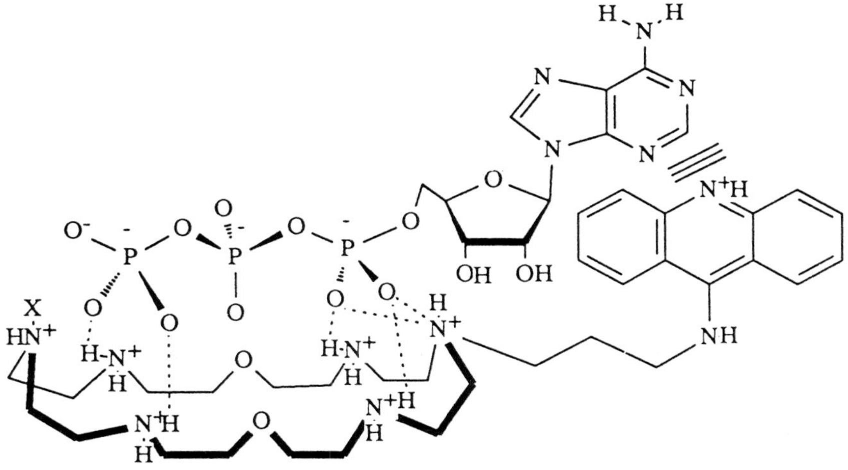

X = H or lone pair of electrons

25

polycation is then capable of multiple hydrogen bond interactions with triphosphate. The sidearm illustrated in **25**, is postulated to stabilize the complex by a charge transfer interaction illustrated by the three lines between aminoacridine and adenine.

Although the structure shown as **25** involves remarkable sophistication, it really is an early example in what will no doubt become an important direction in macrocycle chemistry. Anions are clearly as important as cations, at least biologically speaking. Methods for controlling their availability and interactions will no doubt be of great future interest.

Complexation by crowns and cryptands once meant cation binding. We have shown that even this apparently simple reaction is controlled by numerous variables. Far more complicated and varied complexation processes are now known and interactions between crowns and cryptands and anions, neutral molecules, and even other macrocycles are established. Complexation and the formation of what Lehn calls 'supermolecules' will no doubt be a dominant theme in crown and cryptand chemistry for many years to come.

CHAPTER 4

Structural Aspects of Crowns, Cryptands, and their Complexes

4.1 Introduction

The discussion presented in Chapter 3 should make clear the fact that alkali metal cation complexation is one of the crown ether's most important properties. Complexation of these biologically important cations by neutral crown ethers and cryptands was a remarkable discovery. Complexation of transition and coinage metals was well known and understood but attempts to complex Na^+, K^+, and Ca^{2+} had met with little success although numerous charged complexing agents are known for the latter.

In this chapter, we examine three basic issues. First, what are the structures of uncomplexed macrocyclic ligands? Second, what are the structures of crown complexes? Third, how do cryptands complex cations? The information presented in this chapter derives in large measure from *X*-ray crystal structure analysis. It should be borne in mind that crystal structure determination is as much art as science. In order to obtain a crystal structure, one first has to synthesize the ligand, crystallize the complex, obtain suitable crystals, mount them appropriately, obtain good data, and then solve the structure. If any one of these parts fails, then the entire effort fails. Sometimes crystals may be obtained, but they are flawed in one way or another and will not diffract *X*-rays in a regular enough fashion for analysis. Thus some of the most interesting structures that one might imagine cannot be analyzed by the *X*-ray technique simply because of one or more technical reasons.

4.2 Uncomplexed Crown and Cryptand Ligands

The study of solid state structures of macrocycles by Dunitz, Dobler, Truter, Dalley, Goldberg, Trueblood, Boer, Atwood and many others gave insight into the structural requirements of complexation and the geometric requirements of the free ligands. The 18-crown-6 ligand is the simplest one for which a crystal structure is known. Free 12-crown-4 and

15-crown-5 are both liquids at room temperature, although the melting point of 12-crown-4 is 16 °C. Groth has reported the structure of the former and although no structure is known for the latter, it may ultimately yield to low-temperature analysis. In any event, 18-crown-6 forms a very typical complex and its features are informative. The actual structure of uncomplexed 18-crown-6 is shown schematically in **1** and the idealized D_{3d} geometry of the K^+ complex is shown in **2**. When a cation fills the central void, the methylene groups that occupy the central space of 18-crown-6 rotate outward to create a cavity. In the solid, the cavity is a liability since the empty space destabilizes the crystal. When the cavity is filled by the methylene groups, there is less unoccupied space within and the solid has a higher lattice energy.

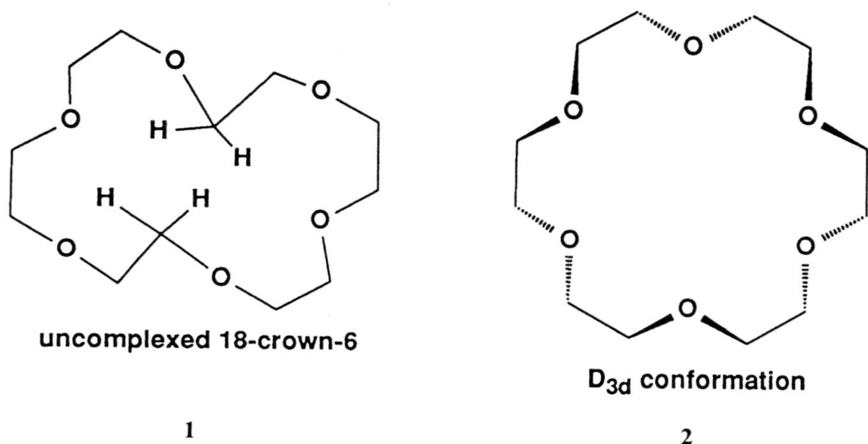

1 **2**

The rotation of a methylene group inward is characteristic of free crown ether ligands and is observed for dibenzo-18-crown-6 and dicyclohexano-18-crown-6. Very few fifteen membered ring structures have been crystallized and subjected to *X*-ray analysis, but in those cases where structures have been obtained, some portion of the macroring rotates inward to fill the molecular void. The macroring of benzo-15-crown-5 is puckered but this ligand has relatively little flexibility. In any benzocrown, the (vicinal) *ortho*-oxygen atoms must be coplanar with the benzene ring. Since the macroring is small, there is relatively little flexibility in the structure.

The Author and his co-workers were fortunate enough to crystallize a cholesteryl derivative of an aza-15-crown-5 compound and the crystal structure was eventually obtained, albeit with some difficulty. The compound is illustrated in **3** and the structure is drawn to show the macroring conformation as observed in the solid state.

Two especially interesting examples are found in the structures of

(a) CPK atomic model of 18-crown-6 in the conformation observed for the uncomplexed ligand

(b) CPK atomic model of 18-crown-6 in the D_{3d} conformation adopted when the ligand binds K^+ (not shown)

dithia-18-crown-6 molecules. When the sulfur atoms replace the oxygens at opposite ends of the molecule, *i.e.*, they are in the 4,13-positions, the sulfur atoms turn outward and the oxygen atoms turn inward to occupy the cavity void as seen in **4**. When the sulfur atoms are separated by only two carbons, the ethylene unit occupies the void as in **5**.

Diaza-18-crown-6 (**6**) and triaza-18-crown-6 (**7**) both adopt the D_{3d} conformation in the free state. The N–H bonds occupy the cavity so no inward rotation of the methylene group is required. The structure of

3

4

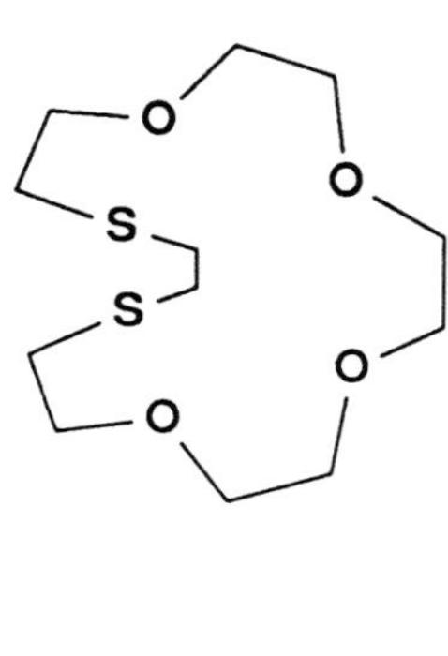

5

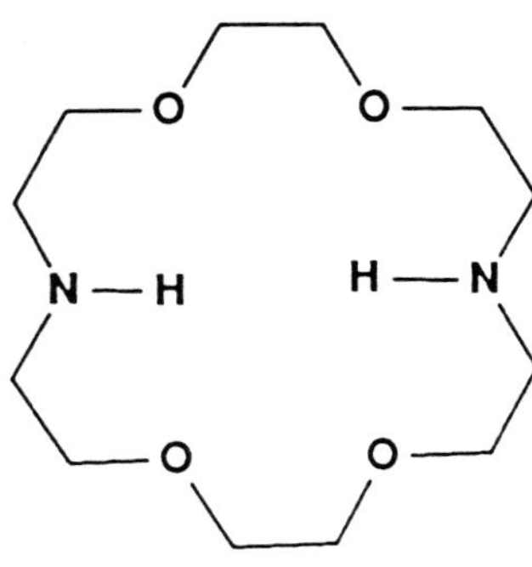

6

7

triaza-18-crown-6 is disordered so that in successive layers of the ligands, oxygen lies atop nitrogen which lies atop oxygen.

Crystal structures have been obtained for the free diaza-18-crown-6 ligand in which the N–H bonds are replaced by *N*-benzyl groups (**8**). The conformation of the ring reverts to that of unsubstituted 18-crown-6. The same is true of *N,N'-bis*(allyl)-4,13-diaza-18-crown-6 (**9**).

The lesson of this section is simple: Nature abhors a vacuum. The empty space that would exist in a crown ether if it adopted a symmetrical conformation with all donors turned inward is energetically unfavorable. As a result, the macrocycles adopt conformations in which the space is filled by a methylene, a hydrogen, or whatever else may be available. We shall see in the next section that a cation readily fills this cavity but that the crown, once again, adopts the conformation that is energetically the best and not necessarily what we might consider the most symmetrical.

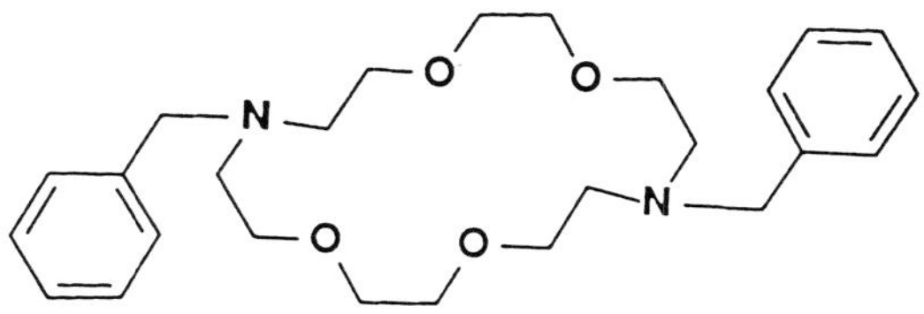

8 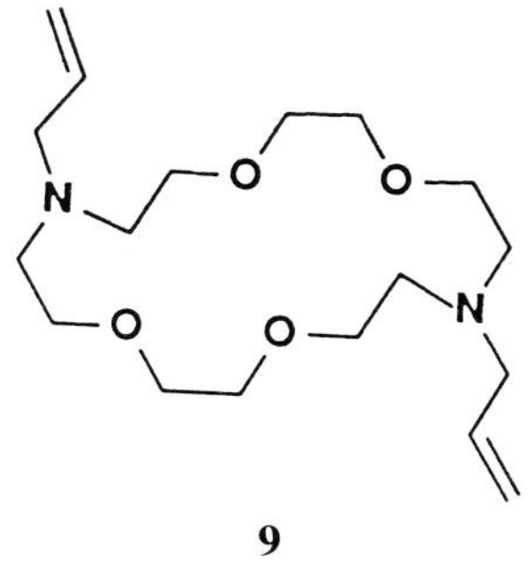 **9**

4.3 Simple Crown Ether Complexes of Metal Cations

Discussions of complex formation have historically been dominated by 'hole' or 'ion-cavity' size concepts. As previously noted, these are not without merit but it would be inappropriate, at the least, to consider that the possibility of forming a complex between a cation and a ligand depended exclusively on size compatibility. Again, the well known and highly symmetrical complex between 18-crown-6 and $K^+(SCN^-)$ has influenced thinking in this area. Its structure is shown in **10**.

The beauty of this structure lies in its symmetry. The idealized crown ether has six-fold symmetry and all K–O bonds are essentially equal in length. Actually, as represented in **10**, three of the oxygen atoms turn upward and three downward. This permits all of the $-OCH_2CH_2O-$ units to adopt the energetically favorable *gauche* conformation. There is also an alternation of direction in the oxygen dipoles which lowers energy. Thus the symmetry of the complex is actually threefold (D_{3d}). The thiocy-

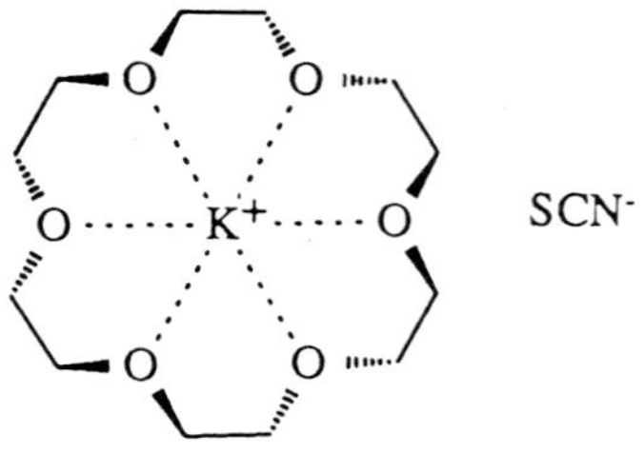

SCN⁻

10

anate anion fills the apical voids in this structure and also provides a counter charge for the complexed cation.

In a way, it is unfortunate that this structure is so beautiful and symmetrical since it has had the historical effect of suggesting that only complexes of idealized size correspondence and symmetry could form. This is certainly not true. Furthermore, those less informed about such structural relationships imagine that the 18-crown-6 adopts the D_{3d} conformation shown even in the absence of a cation. Indeed, a textbook of organic chemistry originating in the United States used a photograph of 18-crown-6 constructed from CPK models for its cover. The uncomplexed crown model has threefold symmetry, a condition never documented in the literature. Still, the picture is a beautiful one. We consider below some of the basic facts about complexation known from solid state structure studies.

If hole size correspondence was critical to complexation, then sodium cation could not be complexed by 12-crown-4 since the latter is too small and 30-crown-10 could not complex potassium. Both of these complexes are known to exist as stable entities. Each complex tells us something different about crown cation interactions but their very existence demonstrates the versatility of these systems.

In many of the structures below, we will not show the actual crystal structure, but the framework structure as shown in Scheme 4.1. Since most crown ether and cryptand complexes have heteroatoms separated by $-CH_2CH_2-$ groups, and since these normally exist in the *gauche* arrangement, an actual structure (ORTEP plot) and the framework structure convey the same information. A CPK molecular model could easily be constructed from either illustration and the exact form of the complex reproduced. Thus the simplified framework illustration at the right in Scheme 4.1 may actually be easier to use in visualizing the relationship of cation and donor atoms.

Three important facts emerge from the data shown in Table 4.1. First, even in the highly symmetrical 18-crown-6 complex (**12**), the bond distances are not identical. In the less symmetrical benzo-15-crown-5 complex (**11**), the differences in M^+–O interaction distances are more pronounced. Second, the Na^+–O distances are shorter than the K^+–O dis-

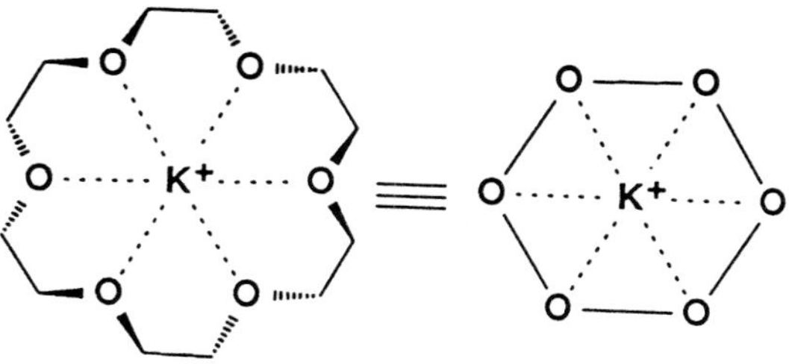

Scheme 4.1

CPK atomic model of the 18-crown-6 · K$^+$ complex. The K$^+$ ion is represented by the silver-colored cube

Table 4.1 *Bond distances (Å) for benzo-15-crown-5 · Na$^+$* (**11**) *and 18-crown-6 · K$^+$* (**12**) *complexes*

M$^+$ · · · O	*Na$^+$ · benzo-15-crown-5*[a] *11*	*K$^+$ · 18-crown-6*[b] *12*
O-1	2.37	2.81
O-2	2.43	2.83
O-3	2.35	2.77
O-4	2.42	2.81
O-5	2.37	2.83
O-6	2.29 (H_2O of hydration)	2.77

[a]I^- counterion. [b]NCS^- counterion.

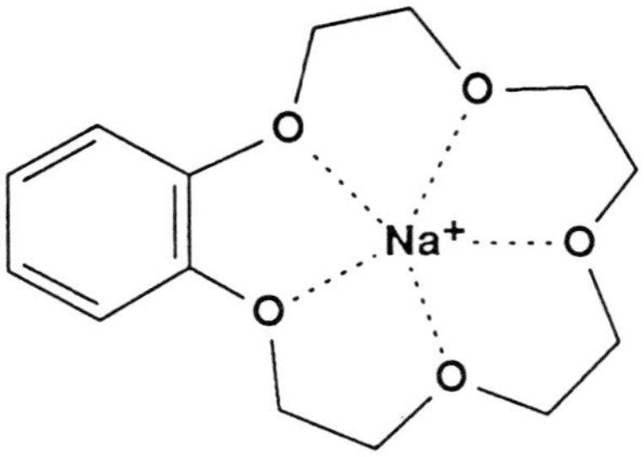

11 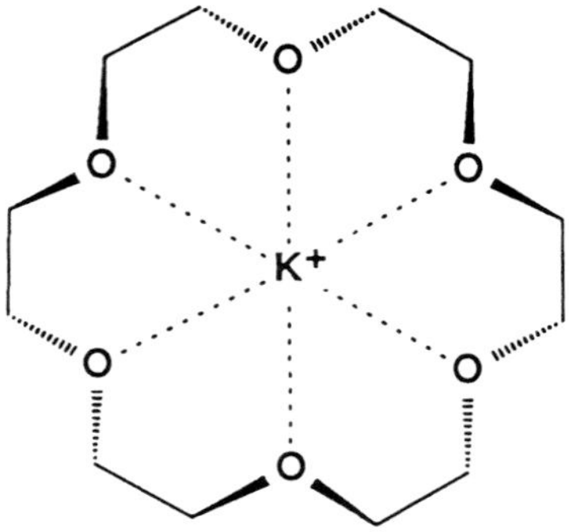 **12**

tances. This is expected but should be noted. We will see later that within a relatively narrow range, these distances vary according to the coordination number of the metal ion. Finally, Na^+ likes to be at least hexa-coordinate but benzo-15-crown-5 cannot provide sufficient donors. Thus a water of hydration is present and this fills the sixth coordination site. In crown complexes, unless the valence shell of the metal is satisfied by heteroatom donors within the macroring, either the counteranion or water will usually be present. In the latter case, the oxygen of water serves as the donor and one of the hydrogen atoms participates in a hydrogen bond to the anion.

4.3.1 Crown Complexes of Cations that Are 'Too Large'

12-Crown-4 (**13**) seems to be too small to accommodate Na^+ since its cavity is only about 1.5 Å at best and the ionic diameter of Na^+ is nearly 2 Å. Another, and probably more serious difficulty is that 12-crown-4 can provide only four donor groups and Na^+ prefers a solvation number of at least six. Both problems are solved by the formation of a sandwich structure of stoichiometry $(crown)_2 \cdot Na^+X^-$ (**15**). Sandwich structures of 12-crown-4 form with a number of cations and Na^+ forms a nearly identical complex (**16**) with aza-12-crown-4 (**14**). In the latter case, the two nitrogen atoms are not on opposite sides as might be expected but separated by an angle of only 43°. This appears to be due to the formation of a long hydrogen bond with the iodide counterion. The latter is illustrated with a top view of the sandwich complex (**17**).

The Na^+–O bond distances for the four oxygen atoms in each ring are 2.47 Å, 2.49 Å, 2.51 Å, and 2.52 Å. Notice that these distances are almost 0.1 Å longer than the bond distances recorded for benzo-15-crown-5 · Na^+ (**11**). In the latter case, Na^+ is coordinated to six oxygen atoms, five from the crown and one from water. In the present case, Na^+ is bound to eight oxygen atoms. The bond or interaction length is reduced as the coordination number is increased because the larger number of

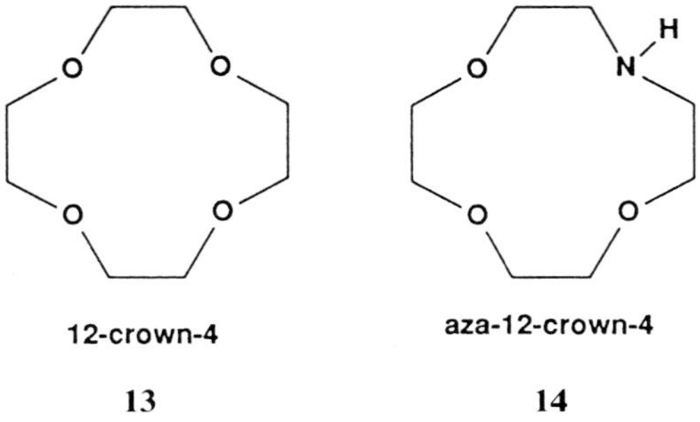

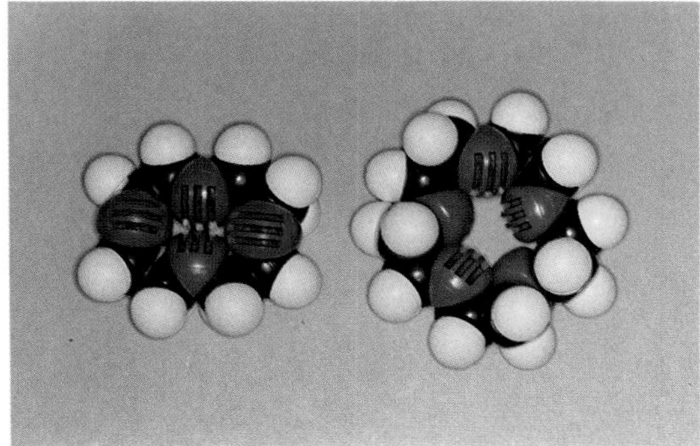

CPK atomic models of 12-crown-4 and 15-crown-5 in conformations that show the maximum cavity size

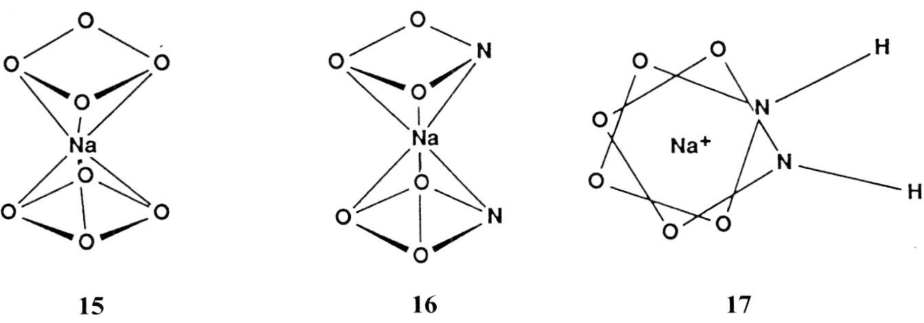

oxygen atoms do not need to interact as strongly with the Na^+ as each oxygen does in the benzo complex where there are fewer donors.

The potassium cation is much larger than the cavity of 15-crown-5. Under no circumstance can the cation fit inside the crown cavity without rupturing a covalent bond. This obviously prevents insertion but that situation would be energetically unfavorable for another reason: K^+ usually favors hexa- or heptacoordination. A 15-crown-5 ligand simply has too few donors as well as being too small to accommodate the cation in its cavity. This analysis is almost identical to that proffered for the 12-crown-4 · Na^+ (**15**) case. As might be expected, the complex formed between benzo-15-crown-5 (**18**) and K^+ is similar to that formed by 12-crown-4 and Na^+. In this case, the cation is eight-coordinate, a common situation for potassium. The ligand (**18**) and the sandwich complex it forms with K^+ (**19**) are shown.

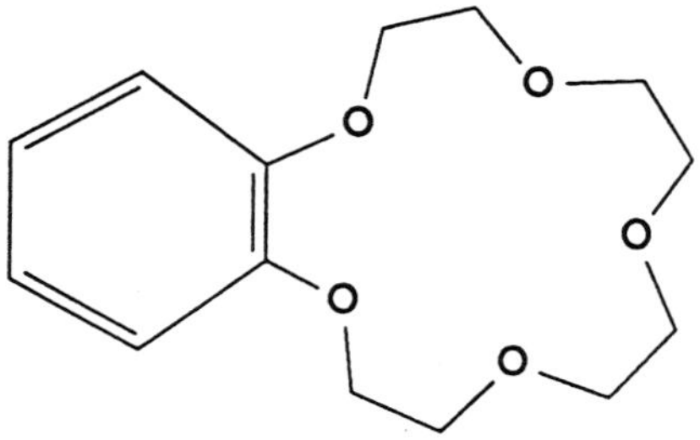

benzo-15-crown-5

18 **19**

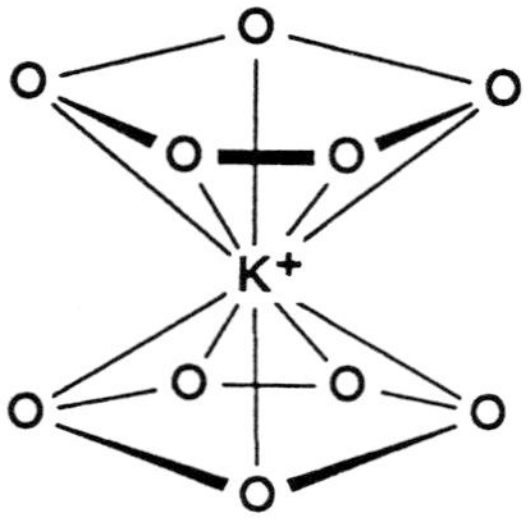

The general idea of cations that are too large for the cavity can be extended even to cesium (Cs). In the case of 18-crown-6 (**20**), however, six donor atoms are available and this is nearly enough to satisfy Cs. Of course, the cation is still too large to be included in the crown cavity, so a sandwich structure of sorts forms. In the $(18\text{-crown-}6)_2 \cdot (CsSCN)_2$ case (**21**), Cs^+ is solvated primarily by a single crown ligand and then shares the sulfur atoms on each of the two counterions. The overall structure may be thought of as a 'half-sandwich' in which Cs is eight-coordinate.

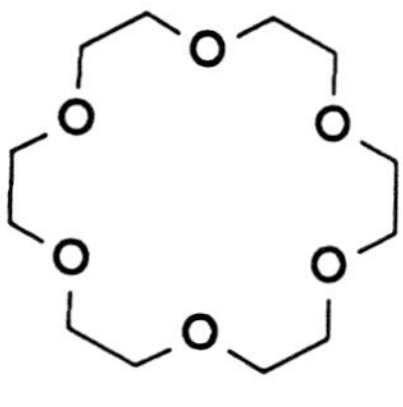

20 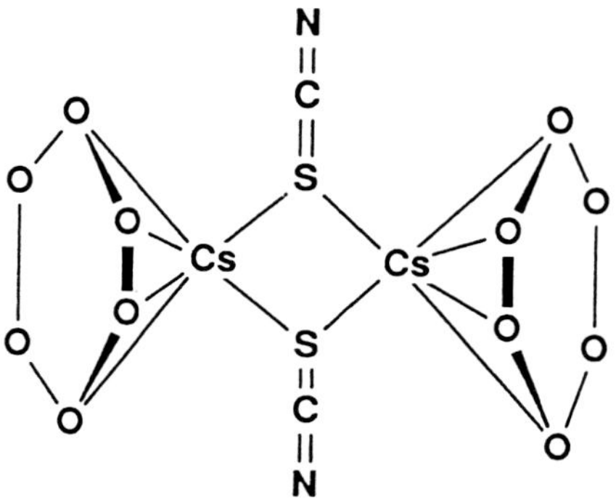 **21**

4.3.2 Crown Complexes of Cations that Are 'Too Small'

If the first major misconception about complexation is that crowns cannot bind cations larger than their cavities, the second misconception is its converse. Crowns readily complex cations that appear to be too small. They do so either by constricting their cavities or, in a few cases, complexing more than one cation. The Na^+NCS^- complex of 18-crown-6 was reported in 1974. It is not the symmetrical structure observed for the K^+ complex but most of its features are similar. The Na^+–O bond distances are, as expected, shorter than the K^+–O bond distances in the 18-crown-6 · K^+ complex (**12**). Five of the six oxygens are involved in a complex quite similar to the 18-crown-6 complex (**22**).

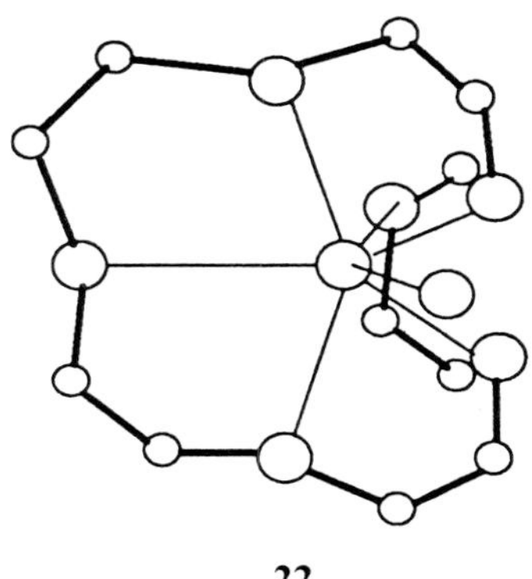

22

A second well-known, but still somewhat unusual example of a complex formed between a large crown and a small cation is the disodium complex of dibenzo-24-crown-8. In this complex, two sodium cations are bound by the 'top' of the macrocycle while the bottom chain extends somewhat to relieve the charge repulsion created within the cavity. Two *ortho*-nitrophenolate counteranions each provide two additional donor sites (nitro and phenoxide) as well as charge neutralization. Indeed, the phenoxide oxygen appears to bridge both sodium cations thus reducing the charge density within the macroring (**23**).

23

Finally, in this brief consideration of crowns that seem to be too large to complex a given cation, we consider a very important example: the dibenzo-30-crown-10 complex of K^+. The parent molecule is large, dwarfing the cation. Nevertheless, the ligand wraps about the cation to provide the latter with ample solvation. Mary Truter has characterized the ligand in this complex as adopting a 'tennis-ball seam' arrangement. Dobler, in his excellent book *Ionophores and their Structures* (p. 166), has pointed out that this structure bears 'a striking similarity to the K^+ complex of nonactin'. He further notes that the mean K^+–O distance of 2.88 Å for this complex is similar to that of nonactin (2.80 Å) and the slightly longer mean length is due to the higher number of donors present (10) in the present complex compared to nonactin (8).

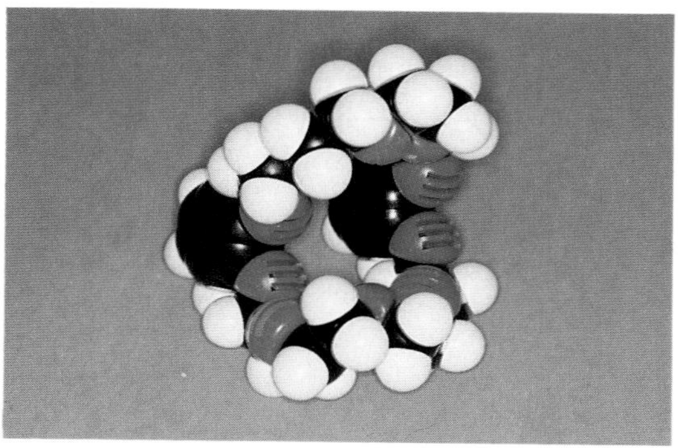

CPK atomic model of the dibenzo-30-crown-10 complex of K^+ in the so called 'tennis ball seam' arrangement

It is also well known that the natural ionophore valinomycin wraps around the K^+ cation and thus transports it *in vivo*. The importance of these observations lies in the fact that transport is best when the carrier can form a three-dimensional complex. The polar cation is well-protected from the lipid environment. A rigid, three-dimensional complexing agent can accomplish the same things but fails in transport due to the poor release rates typical of these complexes (see Chapter 3).

4.3.3 Complexes of Functional Crown Ether Derivatives

Two of the variations in crown ether structure that have engendered greatest study are replacing an ether oxygen either by another function, by another heteroatom, or by incorporating sub-cyclic units. Numerous examples of such systems exist. Two are illustrated (**24** and **25**) here that

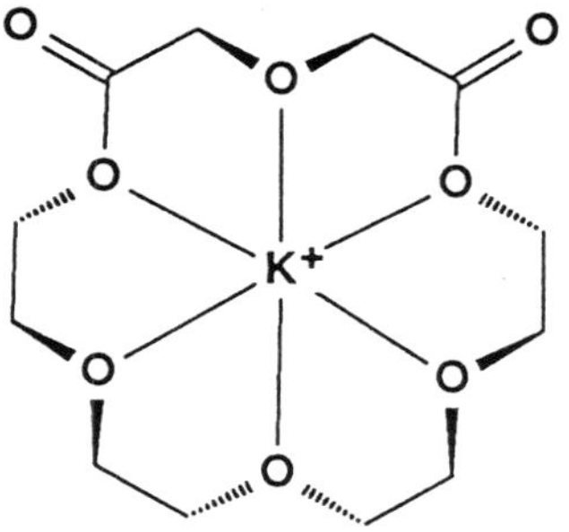

24

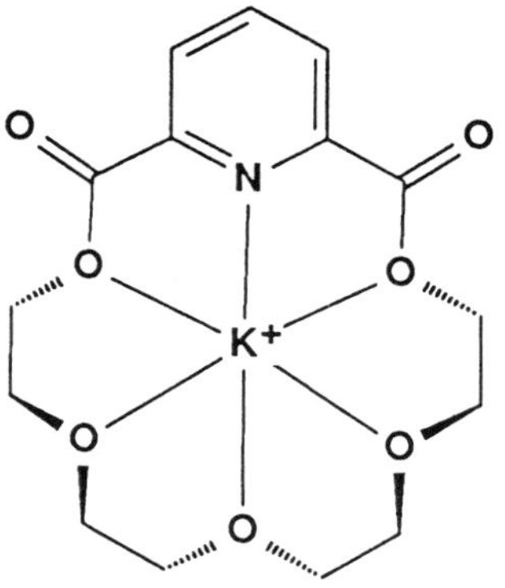

25

show how little overall complex structure may be affected by significant changes in ligand structure. The two examples are both macrocyclic diester compounds, one of which is a derivative of dipicolinic acid (**25**). The potassium cation complexes of these ligands differ remarkably little from that of 18-crown-6.

One of the reasons offered for the preparation of these compounds was to develop valinomycin mimics. The crystal structures show that although these compounds may mimic the binding or transport properties of that remarkable antibiotic, the ester groups are not involved in the cation binding. This fact does not change the structure but points to the problem of variation when more than one potential binding site is present. Nature uses this very wisely in valinomycin where the amides hold the macrocycle's conformation and the ester groups complex the cation. In this case, the ester groups turn away and are not involved in the binding. On the other hand, the free ligands have the virtue of being readily accessible by straightforward synthetic methods.

4.3.4 Lariat Ether Complexes

Efforts in the Author's laboratory have focused to an appreciable extent on crown ethers that have both a traditional macroring and one or more sidearms that contain donor groups and can also interact with a ring-bound cation. As noted previously, these compounds are called lariat ethers because of their ability to 'rope and tie' the cation as well as because their CPK molecular models resemble a looped rope.

The most important finding to emerge from the systematic study of these compounds is that in flexible systems, cations have little preference for one ring size over another. We have already seen that Na^+ forms solid complexes with 12-crown-4 (a sandwich), benzo-15-crown-5, and 18-crown-6. We further systematized these findings by preparing a range of complexes in which the rings and cations were both 'matched' and 'mismatched' in terms of the (macrocycle cavity):(cation diameter) ratio. Consider for example the group of K^+ complexes shown in structures **26–28**. All three of the ligands illustrated have seven donor groups: six oxygen atoms and one nitrogen. They are arranged in quite different ways, however. In the extreme cases, the ring sizes are 12 and 18 atoms and the extreme sidearm lengths are 10 and 3 atoms.

Because the readily invertible nitrogen atom is present at the pivot atom (the point at which the sidearm is attached to the ring), each system is highly flexible. Attachment at one of the carbon atoms would not confer this property upon the ligand as carbon is not invertible.

Potassium cation complexes of each system have been prepared and *X*-ray crystal structure determinations conducted. In these experiments, the cation and the number of donors remain constant. Moreover, each molecule is the isomer of the other so the variation lies completely in the relationship of ring size to sidearm length. Each ligand has the minimum six donors that appears required for complexation on a 1:1 basis with a cation. The structures of the complexes (**26–28**) are shown below in the framework schematic fashion.

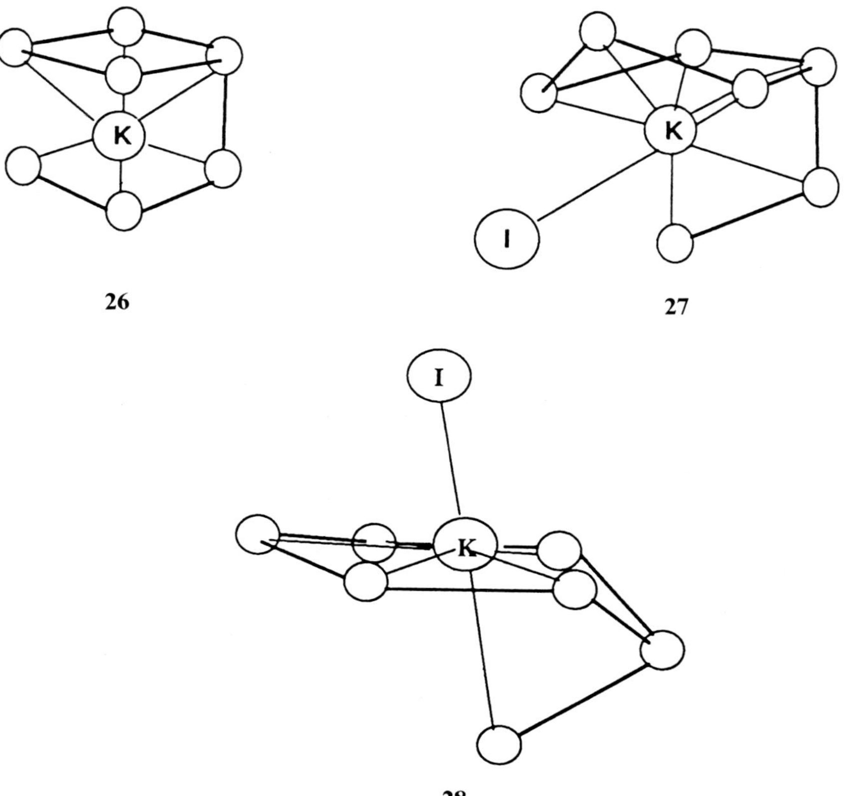

26 **27**

28

It is interesting to note that the potassium cation is slightly above the mean plane of the oxygen atoms. Since the cation requires three-dimensional solvation, it is not surprising that the iodide anion is within the cation's solvation sphere.

Another example from the Author's lariat ether work is informative. It is a 15-membered ring complex of Na^+ in which the sidearm contains an ester donor group (**30**). As with the structure described above, there is a reasonable 'ring size fit' between cation and macrocycle, but the cation is not within the crown cavity. Furthermore, the counteranion is also in the solvation sphere. Two differences distinguish the ester structure from the ether structure above. First, the sidearm contains a carbonyl donor group and this is involved in complexation (see structure **29**). Indeed, the ester oxygen is the closest of all donors to the sodium cation. Second, the counteranion is on the same side as the sodium ion which does not penetrate the macroring cavity nearly as deeply in the present case as in the previous one.

The ether to sodium distances for the complex shown in **30** are: 2.48 Å, 2.51 Å, 2.52 Å, and 2.62 Å. The Na^+–N distance is 2.58 Å. The Na^+ to ester oxygen distance is 2.45 Å suggesting that this is the strongest of all donor group interactions.

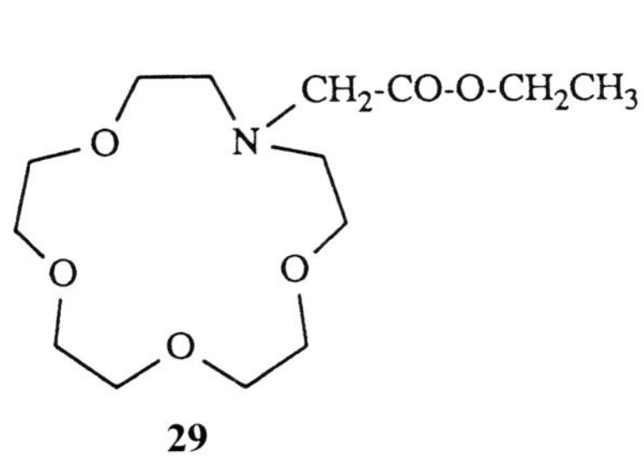

29

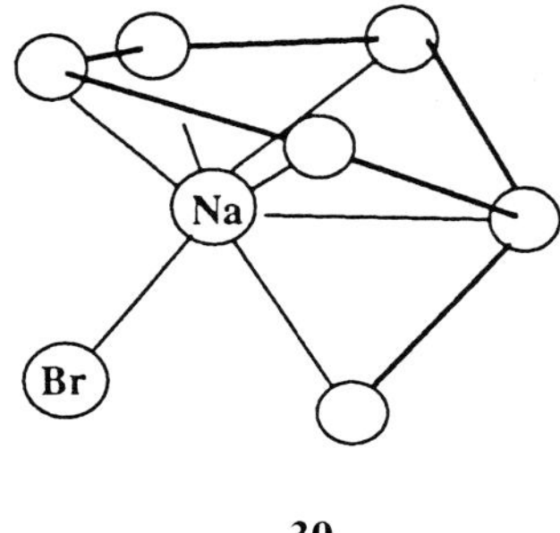

30

4.3.5 Bibracchial Lariat Ether Complexes

We have seen in the previous section that the cation is not necessarily complexed within the macroring and that the counterion may not be in the solvation sphere at all if the crown provides sufficient donor groups. It seems obvious to ask, what geometries will be adopted by two-armed lariat ether complexes? Fortunately, a fairly large sample of crystalline complexes could be obtained and the answer is reasonably clear. Four closely related complexes serve as examples. Two are sodium complexes and two are potassium complexes. In both cases the ring sizes are the same and the sidearms are either 2-hydroxyethyl or 2-methoxyethyl. The free ligands are shown in structures **31** and **32**.

CH_2-CH_2-O-H

N O O O O N

CH_2-CH_2-O-H

31

CH_2-CH_2-O-CH_3

N O O O O N

CH_2-CH_2-O-CH_3

32

Although numerous complex structures are possible, the two most obvious variations are to have the cation stabilized by the sidearm donors from the same side or from opposite sides. We might refer to these as *syn* (same side) and *anti* (opposite sides). As a result of fine work by the Author's co-workers and some good luck, complexes of NaI and KI with each of the ligands were isolated and Fronczek was able to obtain crystal structures of all four. Using the framework style of illustration, we can represent the two principal complex types as shown in **33** and **34**.

When the sidearm was CH_2CH_2OH and the cation (M in the figure) was either Na^+ or K^+, the observed structure was *syn*. When the cation was Na^+ and the sidearm was $CH_2CH_2OCH_3$ (2-methoxyethyl), the complex was in the *syn* arrangement as well. The surprise came when the sidearm was 2-methoxyethyl and the cation was K^+: the complex was in the *anti* conformation. It therefore appeared that the *syn* arrangement, which is very much like the cryptand complexes as we shall see, is the most stable arrangement. The *anti* complex appeared to be an exception. In fact, the exception proved to be the K^+ complex having CH_2CH_2OH sidearms. It now seems clear that Na^+ complexes of 18-membered ring, bibracchial lariat ethers favor the *syn* arrangement and K^+ favors the *anti*

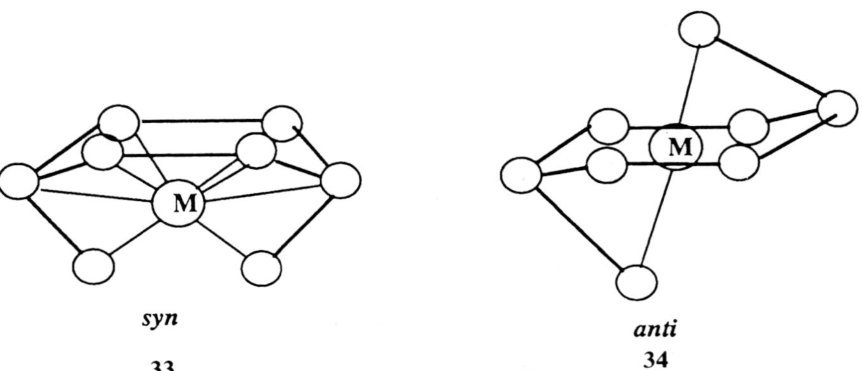

syn

33

anti

34

arrangement. This is because the sidearms of K^+ complexes are too sterically crowded when they are on the same side. The single exception observed so far is for the CH_2CH_2OH group, evidently because H is so small.

4.4 Cryptand Complexes

Generally, crown and cryptand complexation parallel each other. A vast amount of crystal structure work has been undertaken in this area by Moras, Weiss, Metz, Lehn, and others. As with uncomplexed crown ethers, cryptands having no guests compensate for the vacant cavity by rotating a methylene group inward and extending somewhat. The three triethylenedioxy chains that connect each nitrogen are approximately parallel to each other and the N · · · N distance across the molecule extends to nearly 7 Å (**35a** and **35b**).

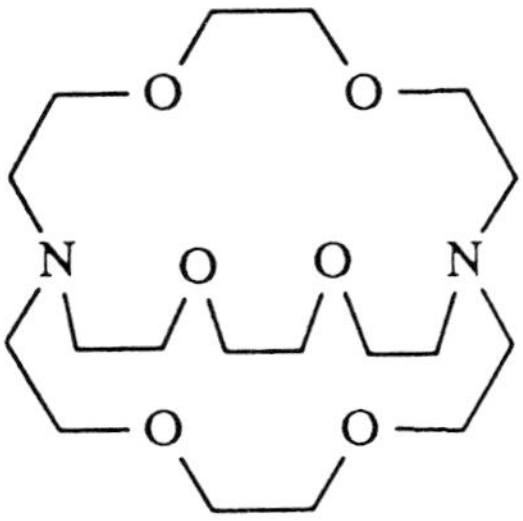

35a

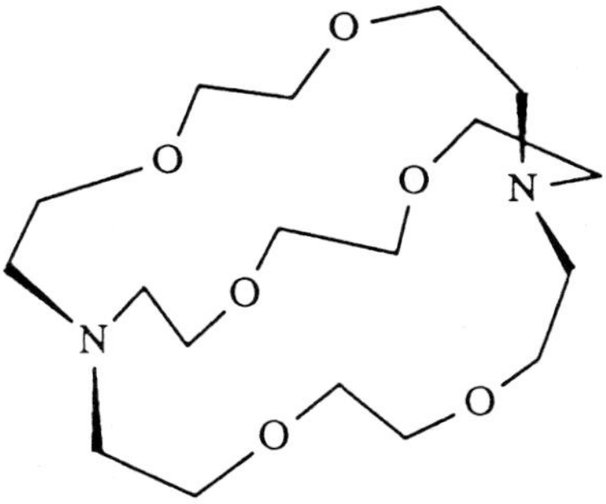

35b

Just as the predominance of structures in the crown area have been reported for 18-crown-6 and its derivatives, the largest number of structures studied are for complexes of [2.2.2]-cryptand (**35**). In accord with Lehn's original suggestion and now extended nomenclature, cryptand complexes are called cryp*tates*.

4.4.1 Simple Cryptand (Cryptate) Complexes

The internal size of [2.2.2]-cryptand (**35**) is about 2.8 Å and this corresponds nicely to the diameter of spherical potassium cation (2.66 Å). The cryptate of [2.2.2] and K^+ is thus relatively unstrained and symmetrical. Most cryptate complexes exhibit approximately three-fold symmetry about the N–M–N axis in which M is the included cation. Dobler has pointed out that the complexation geometry of [2.2.2]-cryptates can be approximated by using the offset triangle schematic often used by inorganic chemists to represent an octahedron. We show in **36a–c** three representations of the $K^+ \cdot$ [2.2.2] complex that suggest its high symmetry.

In the illustrations **36a–c**, the potassium complex of [2.2.2]-cryptand is represented in three ways. **36a** is a simple schematic using the normal line-angle drawings familiar to chemists. **36b** is a schematic view of the complex along the N (represented by a circle)-K^+–N axis. The heavy lines represent the orientation of the complexing ethyleneoxy chains. Dobler's crossed triangle **36c** representation has been adapted in the lower right to show that the three-fold symmetry is not perfect. Indeed, the angle subtended by O4–N–O7 is 22.5°.

The Na^+ complex of [2.2.2]-cryptand is similar, but the fit is not as satisfactory. Thus, there is some contortion of the donors to accommodate the smaller (1.95 Å for Na^+ *vs*. 2.66 Å for K^+) diameter for sodium. A comparison of the two structures is shown in **37** and **38** using framework diagrams.

The Na^+ cation is of appropriate size to fit [2.2.1]-cryptand. Except that one of the chains in the molecule is shorter by an ethyleneoxy unit, the $Na^+ \cdot$ [2.2.1] complex resembles the $K^+ \cdot$ [2.2.2] complex. In Table 4.2 below, bond distances are shown for these two complexes along with the Na^+ complex of [2.2.2], discussed above.

It is interesting to note that although K^+ and Ag^+ are similar in size, the N–Ag^+ bond distances in [2.2.2]-cryptates are significantly shorter than are the corresponding N–K^+ distances. This is in accord with the well known affinity of Ag^+ for nitrogen, especially if two nitrogen atoms are at a 180° angle from each other.

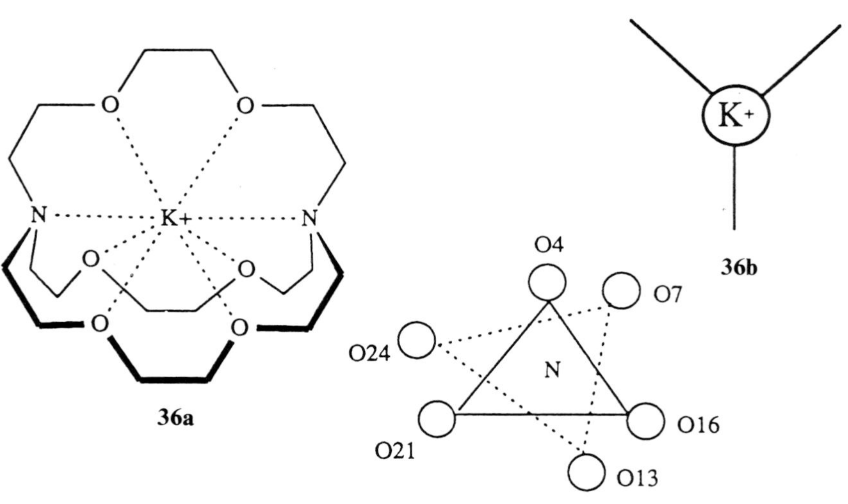

36a **36b** **36c**

Table 4.2 *Comparison of Bond Distances for Cryptate Complexes*

Cryptand	*Cation*	*Avg N· ·M⁺*	*Bond Distance*[a] *Avg O· ·M⁺*	*Range*[b] *M⁺–O*	*Range M⁺–N*
[2.2.1]	Li^+	2.29	2.13	2.29	2.13±0.05
[2.2.1]	Na^+	2.65	2.48	2.48±0.04	2.65±0.06
[2.2.2]	Na^+	2.76	2.58	2.57±0.01	2.75±0.03
[2.2.2]	K^+	2.87	2.79	2.78±0.01	2.87
[2.2.2]	Ag^+	2.48	2.73	2.71±0.05	2.48

[a]In Ångstroms. [b]If no variance is given, both or all bond lengths are the same

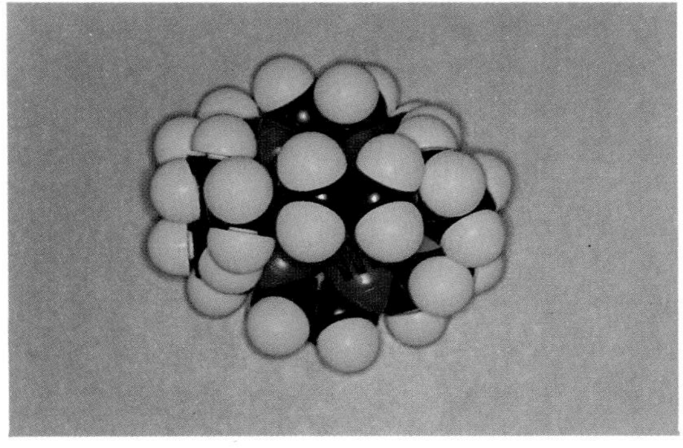

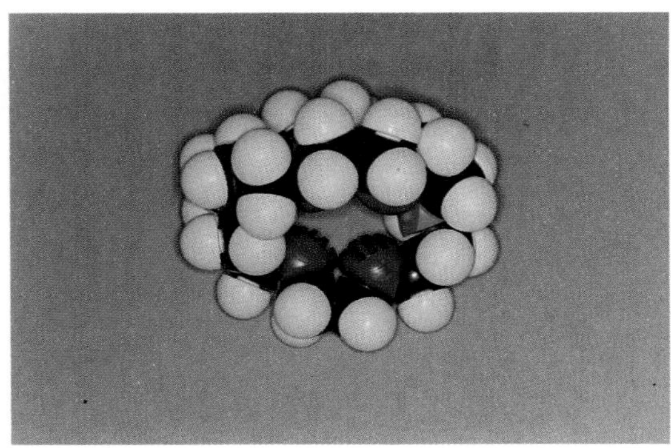

CPK atomic model of [2.2.2]-cryptand in the binding conformation but without a cation (a) top view and (b) side view. In the absence of a cation, the cavity is filled by inward-turned CH_2 groups as with crowns

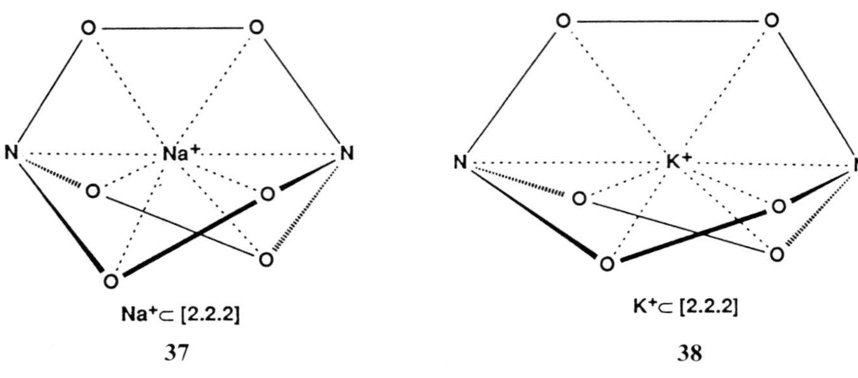

$Na^+ \subset [2.2.2]$

37

$K^+ \subset [2.2.2]$

38

4.4.2 Ditopic Cryptands

Macrotricyclic or ditopic receptors have two crown ether rings that are bound together. A number of such structures are known as are their complexes. The ligand having two 12-membered rings separated by a diethyleneoxy unit accommodates two silver cations (**39**) within it and the corresponding *bis*(diaza-18-crown-6) system complexes two sodium cations (**40**). In the latter case, each Na^+ is solvated by the four oxygen atoms and both nitrogen atoms in the ring as well as by one of the spacer oxygens. Thus each Na^+ is seven-coordinate and the complex is 'tilted' so that the oxygen atoms in each chain can solvate opposite macroring-

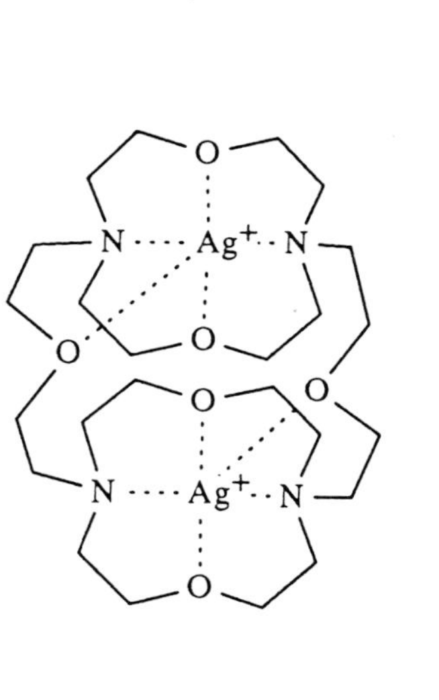

39

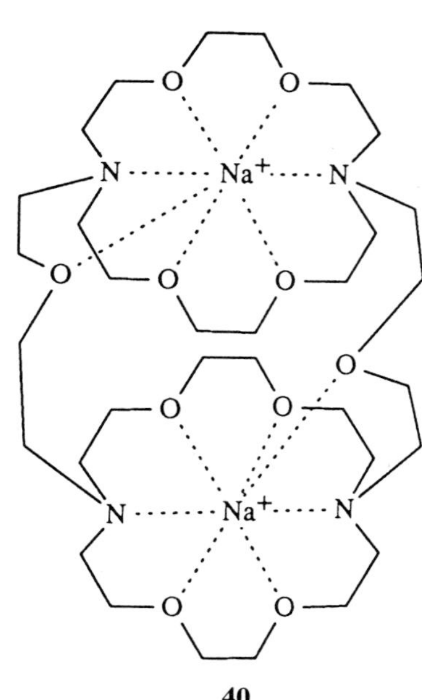

40

CPK atomic model of a ditopic cryptand comprised of two diaza-18-crown-6 rings connected at nitrogen by CH_2-C≡C-C≡C-CH_2 bridges (cations not shown)

bound cations. The silver cation is likewise solvated by the four donors in each macroring and one of the two available spacer oxygens.

It might seem unusual that the smaller ligand accommodates the larger cation rather than *vice versa*. We must recall that silver has a high affinity for *trans* or opposite nitrogen atoms and the relatively short bond lengths reflect this preference.

4.4.3 Inclusion *vs*. Exclusion Complexes

We have discussed (Sections 4.3.1 and 4.3.2) the numerous complexes of crown ethers that do not correspond in size to the cation bound. When the cation seems too large, two crowns often participate in a sandwich complex. When the cation seems to be too small, more than one cation may be bound by the ligand, or the ligand may wrap about the cation to better accommodate the required bond lengths and coordination number. There is a somewhat different situation with cryptands since they are far less flexible (more rigid) than the 'two-dimensional' crown ethers. Some cations are simply too large to fit inside the cryptand cavity. When this occurs, an 'exclusive' rather than 'inclusive' complex may form. All of the complexes illustrated previously are of the inclusive type. There are two exclusive examples that deserve mention here.

The first is the simple case of $K^+ \cdot$ [2.2.1] (**42**). The ligand is too small for the cation to fully penetrate into the cavity. Binding therefore occurs on the 18-membered ring surface of the cryptand and the cation is solvated from above by the short-chain-oxygen. The anion, thiocyanate in this case, solvates the cation from the bottom. The complexes **41** and **42** are thus quite similar to the lariat ether complex of K^+ shown schematically in section 4.3.4.

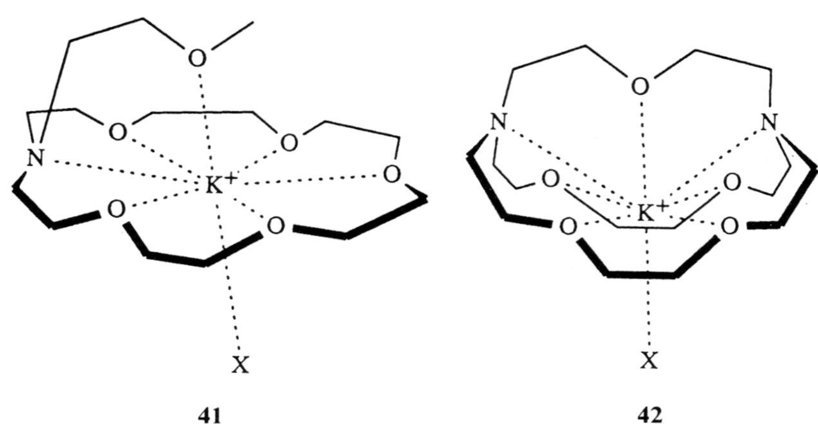

41 **42**

The other special case of an exclusive complex is closely related to the structure shown in section 4.2 for the uncomplexed disulfur macrocycle. In this case, the conformational preference of the two sulfur atoms is to face outward from the macroring. This geometry is adopted also by the [2.2.2] derivative containing two sulfur atoms in one of the chains when bound to palladium (**43**). The palladium dichloride ($PdCl_2$) complex is shown schematically in **43**. Note that Pd is tetravalent and the 18-membered diaza-18-crown-6 macroring is not involved in binding at all.

4.5 Crown Ether Complexes of Other Ions and Molecules

Cram's interest in ammonium ion resolution by using chiral crown ethers engendered an extensive study of ammonium ion–macrocycle interactions. Numerous structures obtained by Trueblood and co-workers on complexes prepared by Cram and co-workers have demonstrated that crown ethers of appropriate size bind ammonium ions in a three-point fashion. An excellent example is the complex of *tert*-butylammonium tetrafluoroborate with 2,3-naphtho-18-crown-6. The complexing elements (**44** and **45**) and the complex (**46**) are shown schematically.

The positively-charged nitrogen atom is above the plane containing the oxygen atoms and forms a tripod oriented as shown in structure **46**. The N^+-$C(CH_3)_3$ bond is only 1° from perpendicularity with the oxygen atom plane and the methyl groups are staggered with respect to the O · · · H–N bonds. As seen in previous structures, all of the macrocycle's $-CH_2CH_2-$ units are *gauche*. The tetrafluoroborate anion is well outside the solvation sphere of ammonium.

Cram and co-workers also reported what must be the most extreme example of hydrogen bonding in a simple crown–ammonium complex. The structure of 18-crown-6 complexing monoprotonated hydrazine

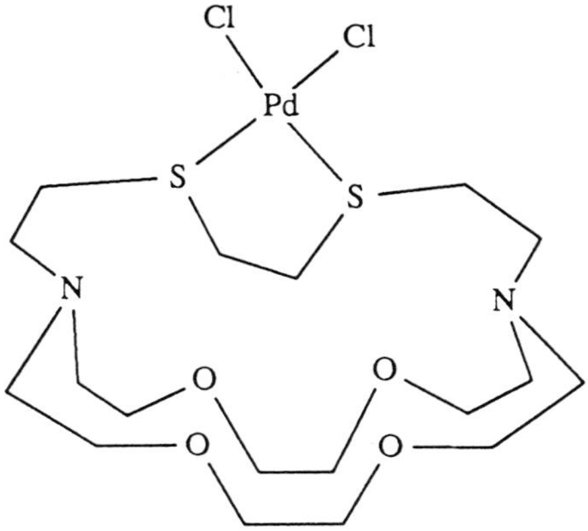

43

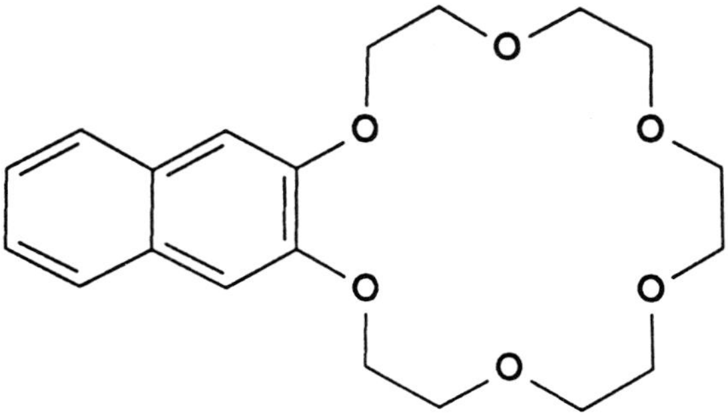

44

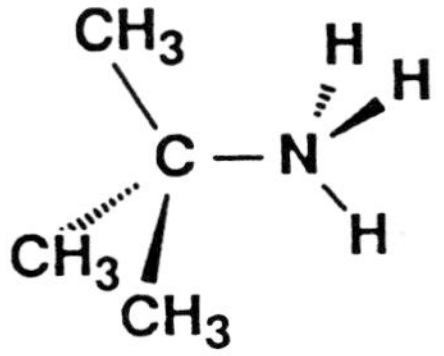

45

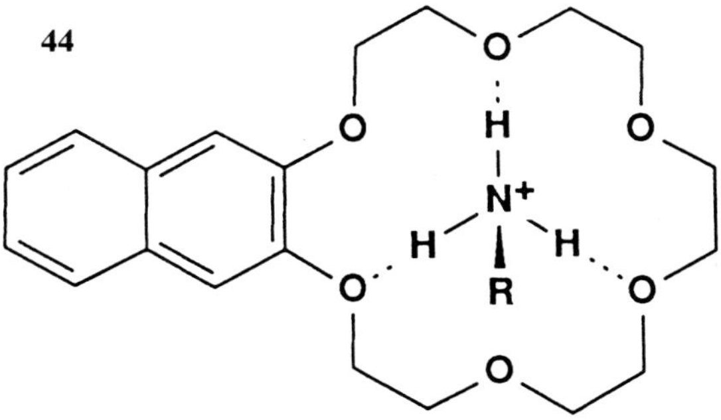

46

involves five hydrogen bonds. The H–N–N–H dihedral angle is 60°. The three N^+–H · · · O bond distances have an average value of 2.84 Å and the two N–H · · · O distances are 3.05 Å and less linear than the former. The structure is shown schematically in **47**.

When a 2,6-pyridyl unit is present in the 18-crown-6 macroring, an ammonium cation complex is still possible, but the structure is signifi-

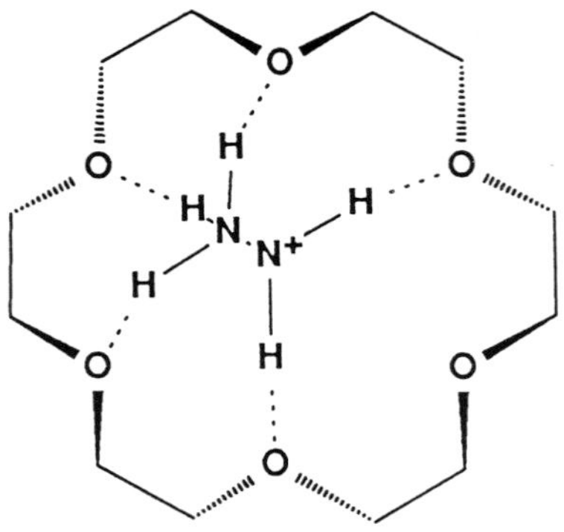

47

cantly altered. Since the pyridyl nitrogen is more basic than any oxygen in the macroring, it is expected to be involved in binding the ammonium ion. It is, however, sp^2 hybridized so the lone pair is in the plane of the pyridine ring. Either the macroring must distort to accommodate the ammonium ion, or the $-N^+H_3$ tripod must flatten to accommodate the macrocycle. The complex, shown schematically in **48**, has a distorted macroring. Three hydrogen bonds are formed between the ammonium cation and the ring, but they are not quite linear (173°, 175°, and 175°). The N^+–H · · · O distances average 2.13 Å.

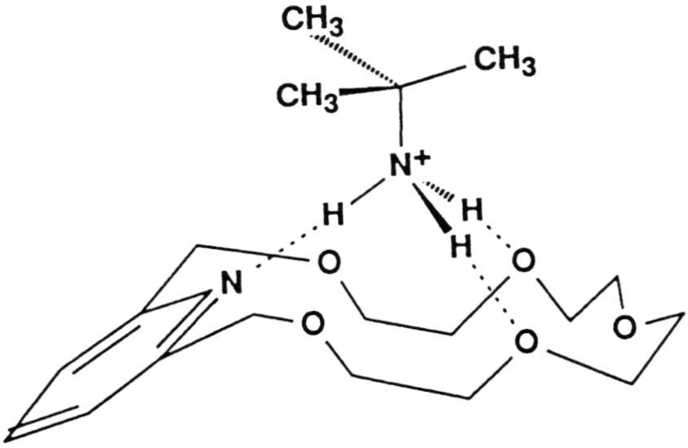

48

A variation on this theme is the hydrated complex of protonated aza-18-crown-6 (**49**). The N–O distance to the water oxygen was found to be 2.727 Å. The O–H–O distances from the water to the peripheral oxygen atoms were 2.898 Å and 2.909 Å, distances that clearly indicate strong hydrogen bonding interactions. Further, the N–H distance is 0.973 Å. This short distance suggests that the complex is a hydrated ammonium

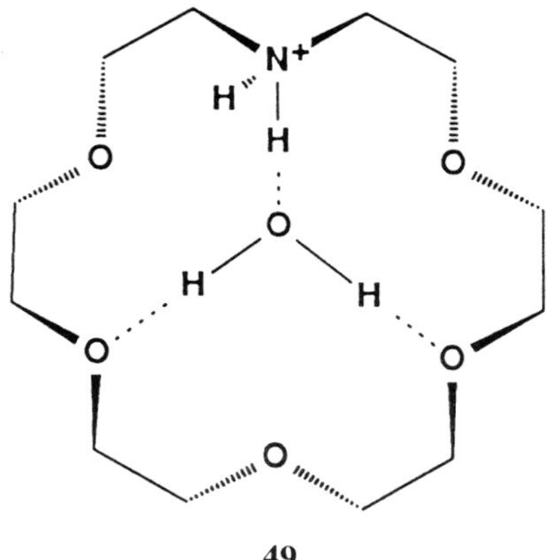

49

complex rather than hydronium complex. The fact that hydrogen bonds are often double well potentials means that precise assignment of structure is unclear.

Aza-18-crown-6 (**50**) was first prepared in the hope that it might form an unusual potassium complex when the free amine was treated with KH. The complex would be of a powerful base but one that was overall neutral because the cation was complexed within the macroring (Scheme 4.2). It is interesting to note that since the crystallographer was expecting to find K^+, the preliminary structure of the compound above showed K^+ above and outside the macroring near nitrogen. Of course, K^+ and Cl^- have the same electronic structure and their scattering factors are similar. Further, the molecular weights of the two complexes shown differ by only 2 daltons and their percentages of C, H, and N are within the normal experimental error of combustion analysis. This story is related because it shows how dangerous simple assumptions can be and that even *X*-ray crystallography is not infallible.

The search for the corresponding hydronium ion complex began shortly after the discovery of crowns and the first example was reported by the Izatt–Christensen group in 1972. Good evidence for the complex was presented but no structure was obtained. Among the questions of

50

Scheme 4.2

interest were (i) whether the complex was completely symmetrical and (ii) whether the hydronium ion was planar as required by sp^3 hybridization.

Several structures have been reported. The French group of Behr, Dumas, and Moras reported a complex between tetracarboxy-18-crown-6 and hydronium that suggested a pyramidal geometry for H_3O^+. More recent work by Shoemaker and co-workers and by Atwood and co-workers suggested that the hydronium cation was closer to planarity. The Atwood structure is especially interesting because it was obtained by bubbling HCl gas through a suspension of 18-crown-6 in moist toluene. The complex $[H_3O^+ \cdot 18\text{-crown-}6][Cl{-}H{-}Cl^-]$ resulted and its structure is shown schematically in **51**. Atwood has studied the interactions of salts and hydrocarbons and dubbed their special interactions the 'liquid clathrate effect'. The planarity of H_3O^+ was judged by the fact that the oxygen atom in the hole was less than 0.1 Å from the plane formed by the three nearest oxygen atoms. In the Moras structure, oxygen lay 0.61 Å from the six-oxygen plane. The Atwood structure **51** is shown without the hydronium oxygen atoms that could not be located although the $O_{water} \cdots O_{ring}$ distances all lay between 2.70 Å and 2.85 Å.

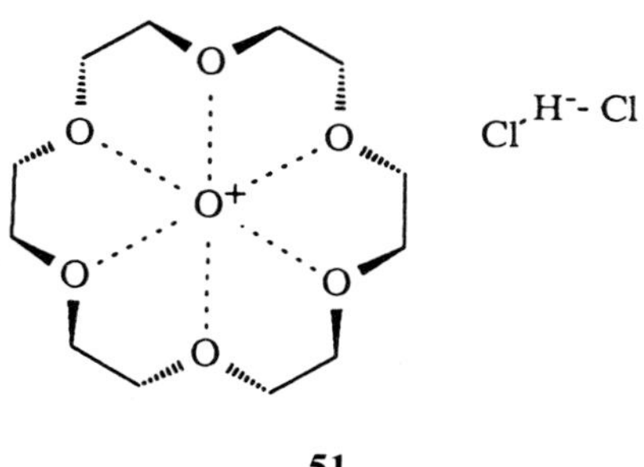

51

Two very interesting complexes that also show intramolecular hydrogen bonding are shown in **52** and **53**. Both are derivatives of 2,6-dimethylbenzoic acid. The two are identical except that the second crown has a transannular pyridine residue while the first lacks it. Their structures are quite similar, however. In both cases, the distance from the carboxyl hydrogen to the opposite atom is 1.8 Å and in both cases it appears that the oxygen atoms flanking the benzene ring interact with the partially positive carboxyl carbon. The O–C bond distances in the benzo-18-crown-5 derivative lacking the pyridine are 2.75 Å and 2.77 Å while these distances are slightly longer at 2.84 Å and 2.87 Å when pyridine is present. Note that the normal van der Waals distance is about 3.1 Å so these short distances are significant. In both cases, the carboxyl is rotated away from planarity with the aromatic ring, an arrangement that is electronically poor but sterically favorable. There is likely some 'no-bond' resonance which stabilizes the carboxyl group. The structures are illus-

52 **53**

trated using standard chemical structures but note that in each case the carboxyl hydrogen bonds are linear.

Another interesting complex is that of water and a lithium cation within the same macroring (**54**). In this case, the cation is too small for the fairly rigid spiro-macrocycle and the cavity is filled by a water molecule. Lithium is also solvated by a second water molecule in each case.

Lehn's remarkable spherand complexes ammonium ion. That this should be so is not surprising since the internal cavity has a tetrahedral array of donor groups. Average hydrogen bond distances in the ammonium iodide complex are 3.11 Å. An even more interesting complex involving this ligand and chloride anion has been reported by Moras and Weiss (**55**). All four nitrogen atoms are protonated and this forms a matrix in which chloride anion can be trapped as shown in structure **55**. The hydrogen atoms which bridge from N^+ to Cl^- are not shown. The average Cl · · · N distances are 3.09 Å and the Cl · · · O distances are slightly longer at 3.25 Å.

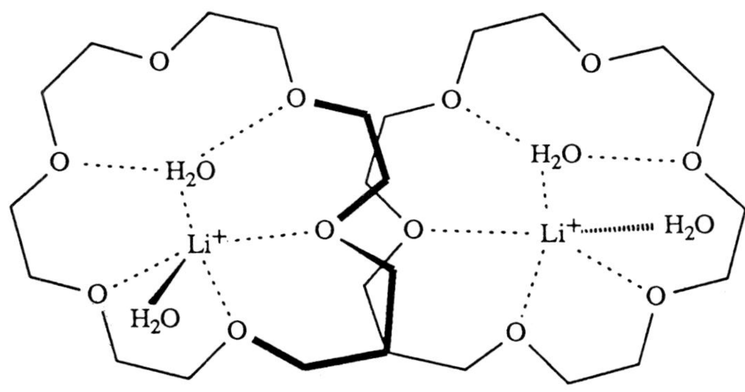

54

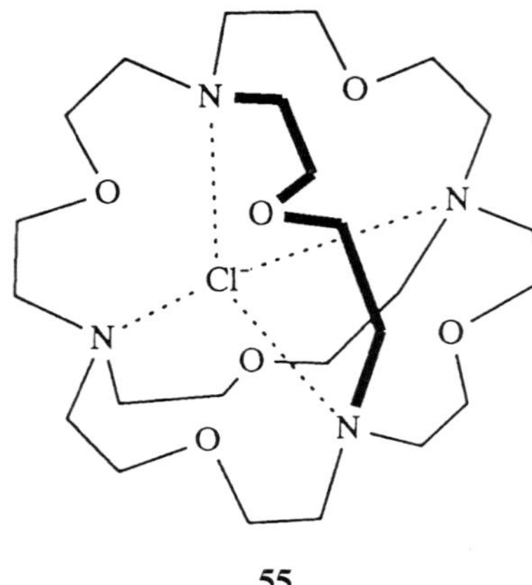

55

4.6 Podand Complexes

While formally beyond the scope of this book, it is certainly in order to mention the extensive studies of podands (non-cyclic, crown-ether-like molecules). The podands range from the obvious non-cyclic, relative of 18-crown-6, pentaethylene glycol dimethyl ether, to much more imaginative, open-chained derivatives that are virtually crown ethers or cryptands (tripod ligands) save for cyclization. These systems have been of particular interest to Vögtle and Weber and much of the structural work has been done by Saenger (Berlin).

Probably the most important aspect of these studies is the so-called 'end-group' concept. In essence, this is the recognition that when conformationally flexible systems are rigidified, they are stronger complexing agents and form more ordered crystals. In addition, when donor groups are present in these end groups, the complex is further stabilized and rigidified. It is thus possible to obtain detailed structural information about these complexes even though they are quite flexible and presumably disordered in the non-complexed state.

Two of the end groups that have proved most successful are 2-methoxyphenoxy and 8-quinolinyl. Two complexes are shown (**56** and **57**) which illustrate both the similarities and differences observed in these complexes. The quinolinyl units dominate the structure **56** which is essentially planar. The 2-methoxyphenoxy end groups, on the other hand, twist to provide more three-dimensional solvation of the cation (**57**). In the quinolinyl case, two counteranions (in this case, iodide) provide the third dimension of solvation.

Similar structures have been prepared in which a third chain is present. In the two examples **58** and **59**, the additional chain is connected at nitrogen. Complexes of these with a variety of cations including barium

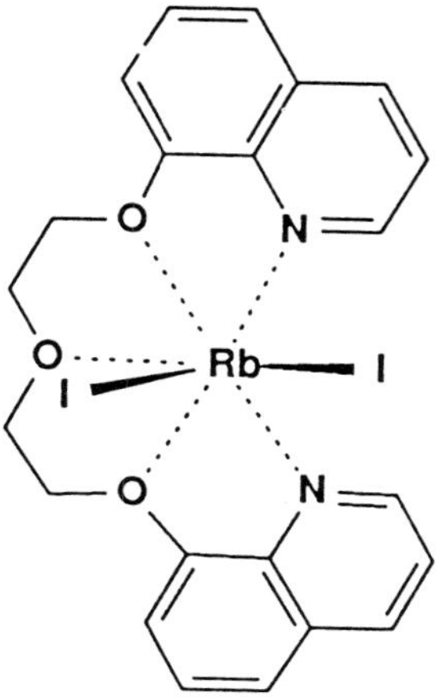

56

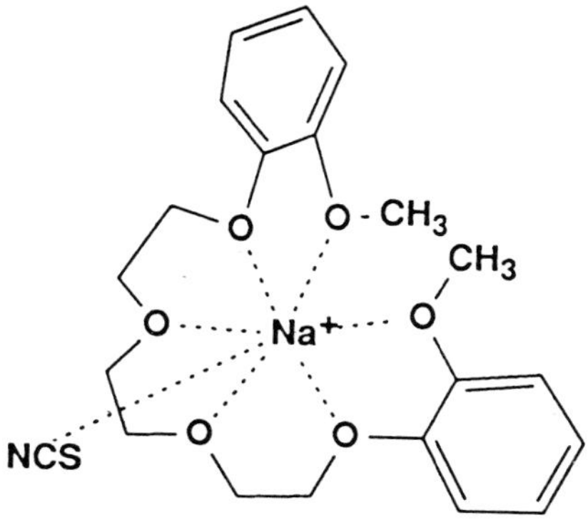

57

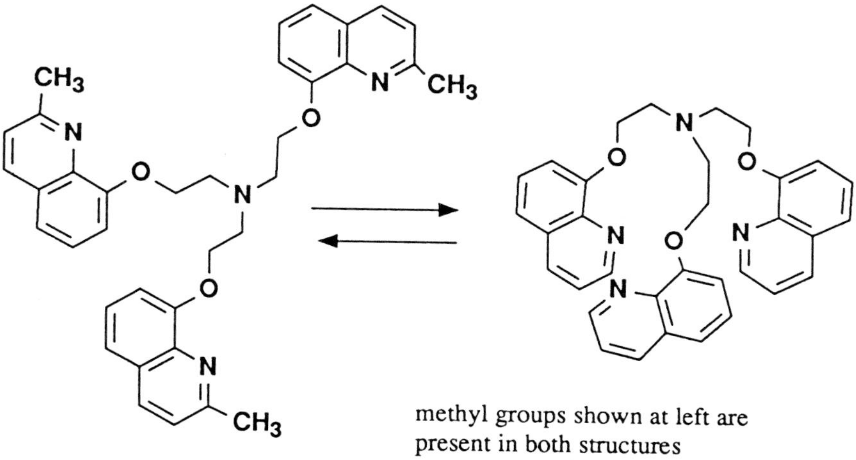

methyl groups shown at left are present in both structures

58

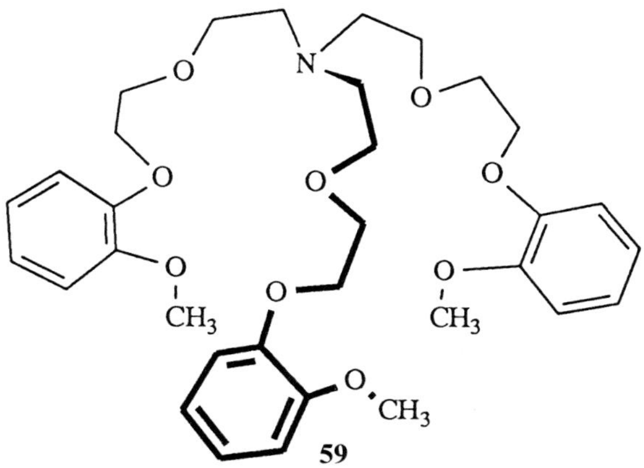

59

and potassium have been studied by Saenger and his associates. Saenger has called these systems 'three-legged octopus' molecules and 'tripodands' both of which seem accurate although the former is certainly more evocative. Vögtle has referred to the tripodands as 'non-cyclic cryptands'.

The potassium thiocyanate complex of the *tris*(anisyl) ligand **(59)** has been studied. The potassium cation is ten-coordinate and the counter-anion is excluded from the solvation sphere. Thus the non-cyclic analogy seems appropriate.

Two features of complexation should now be clear. Crowns and cryptands use donor atoms to stabilize cations or other dipolar species when complexation occurs. Nitrogen or oxygen will usually serve as a donor to a positively charged or polarized guest but protonated nitrogen or other Lewis acid may coordinate to a negative dipole of charge. Second, crowns and cryptands are capable of considerable structural adjustment so that a variety of guest species can be accommodated even when size and shape correspondence is not optimal.

CHAPTER 5

Applications of Crowns and Cryptands

5.1 Introduction

The application of crown ethers and cryptands to chemical processes is of major consequence today and will only increase in importance. It is hard to think that thirty years ago, these compounds were unknown. Indeed, for nearly a decade after their discovery, they were still widely regarded as toxic curiosities.

The presumed toxicity of crown ethers is worth some mention as it shows that even scientists sometimes believe rumors. The myth that crown ethers are dangerously toxic was based on little data and some imagination. Toxicity studies were conducted at an early date on 12-crown-4 at the Dow Chemical Company. Further, Pedersen had noted in the *Organic Syntheses* preparation the unpleasant effect on a dog's eyes when they were exposed to dicyclohexano-18-crown-6. Pedersen reported an approximate lethal dose for dicyclohexano-18-crown-6 for ingestion by rats of 300 mg kg^{-1}. He also reported that a dose of 200 mg kg^{-1} was not lethal in 14 days. Even so, it was not uncommon in the early days to hear purely anecdotal information related at specialist meetings. For example, a scientist reported during a plenary lecture that one or more co-workers had suffered repeated colds or sore throats during the winter. This was attributed to working with certain crown ethers rather than to the winter cold season. No physician was reported to have concurred in this analysis at the time. Another scientist reported feeling depressed while working with a particular crown. Again no qualified medical opinion was presented.

There is always the possibility that these two cases were examples of true physiological effects but to the Author's knowledge, neither story has ever been investigated nor corroborated. Unfortunately, the memory of such reports lingers. It is now known that the macrocyclic polyethers other than 12-crown-4, generally do not exhibit either great toxicity or carcinogenicity, at least among those examples that have been tested. Izatt, has reported that the LD_{50} (the lethal dose to 50% of the test animals) for 18-crown-6 and for aspirin are about the same. This is not to say that crown ethers and cryptands exhibit no physiological activity.

One important potential application of crowns was under study during the early to mid-1970s, at Dupont. It was found that certain benzo-15-crown-5 oxime derivatives exhibited anti-viral activity. This anti-rhinovirus activity might have led to a cure for the common cold. Alas, the crown ethers were metabolized too quickly to build up sufficient concentrations in the sinuses to have the hoped-for therapeutic effect.

Obviously, crowns and cryptands bind alkali metal cations and the latter are critical to many biological processes. Compounds that bind these cations should always be presumed to have biological activity and should be handled accordingly. It goes without saying that all chemicals should be treated with respect, but it now appears that 18-crown-6 requires no more or less vigilance than other common substances. The Author believes that industrial interest in and the general application of crown ethers was retarded significantly because of these unfortunate, unscientific, and unsubstantiated anecdotes.

All of the above is by way of saying that the crown ethers probably were less studied and less utilized, especially in industry, than their potential demanded. As we saw in Chapter 2, there are now a number of methods available for the synthesis of crown ethers. The expense issue has largely been set aside and the value of crown ethers and cryptands is obvious, especially in high technology applications.

In this chapter we explore some of the applications and developments involving crown ethers. As with other Chapters in this short book, these examples can hardly be exhaustive. The eclectic discussion that follows attempts to represent the diversity of applications and properties in the hope that some new ideas will be stimulated. We begin with a discussion of some of the first observations of their properties and then proceed to some of the ways in which they have been elaborated into complex supramolecular systems.

5.2 Solubilization Phenomena

It is an adage in chemistry that 'water is the universal solvent.' This is so because much early study in chemistry was focused on what we would now call inorganic salts. Substances like NaCl or KBr exist by virtue of non-covalent (ionic) bonds. In the crystal lattice of sodium chloride, each sodium cation is surrounded by six chloride anions and each chloride anion is surrounded by six sodium cations. The symmetry of this system is important because each cation is separated from every other cation by a screen of anions. Moreover, any charged species is inherently unstable and the presence of anions stabilizes a cation and *vice versa*.

When sodium chloride is contacted by water, the sodium ion is surrounded by the electron pairs of the water oxygen atom as illustrated in **1**. Likewise the chloride anion is hydrogen bonded by the water

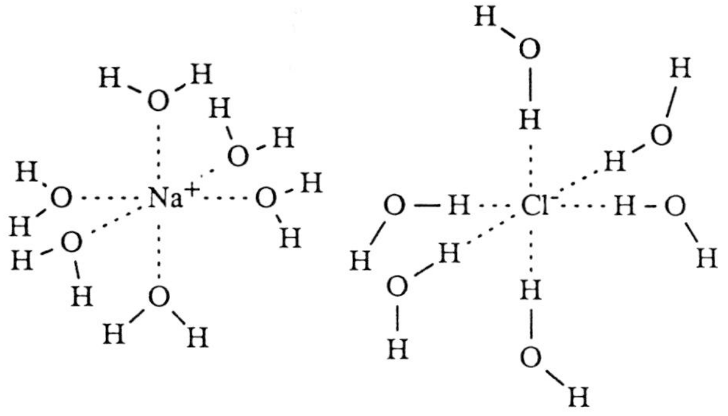

1

hydrogens. This is not exactly the same situation that is observed in the solid but it is conceptually similar. In polar solution, each cation is surrounded by negative dipoles and each anion is hydrogen bonded or is surrounded by positively polarized atoms. Water is uniquely able to stabilize both positive and negative charges. The arrangement of water about Na^+ and Cl^- is called the first solvation shell or sphere.

A macrocyclic polyether such as 18-crown-6 can surround a cation in much the same way. The 18-crown-6 · K^+Cl^- and the K^+ complex of [2.2.2]-cryptand are shown in Chapter 4. In both cases, K^+ is solvated by an array of oxygen (or nitrogen) dipoles. In the case of cryptands, the 'solvation shell' is three-dimensional as expected in water but only two-dimensional in the crown case. Thus the anion will be well away from the cation in the cryptand complexes but directly interacting in the crown case (see Chapter 4).

There are two important consequences of these structures. First, in both cases, the ligand forms a 'solvation shell' about the cation. Thus, K^+X^- complexes are generally not soluble in non-polar solvents such as chloroform, but the crown and cryptand complexes usually are. The inward focused (convergent) oxygen atoms solvate the cation. The hydrophobic exterior is solvated by chloroform so the ligand actually provides a unimolecular solvation shell for the cation.

Second, the fact that nearly all of the solvation is related to the cation and not to the anion means that the anion is reactive relative to its situations in water. Crown ethers typically solvate a cation in two dimensions so there is often a direct $M^+ \cdots X^-$ interaction. Even so, the stabilization of the anion is far poorer than in water and the anion is said to be 'activated'. The extent of the activation is even greater in the cryptate complexes than in coronate complexes because in the former case, the ligand prevents direct interactions between cation and anion. The ability of crown ethers and cryptands to solubilize salts in non-polar solvents and to activate anions has been used to greatest advantage in the process known as phase transfer catalysis.

5.2.1 Solubilization of Salts

Pedersen recognized that potassium hydroxide (KOH) has little solubility in hydrocarbon media. Using the hydrophobic and hydrocarbon soluble macrocycle derivative dicyclohexano-18-crown-6, Pedersen demonstrated these principles by dissolving potassium hydroxide in toluene. He then used the potassium hydroxide complex in toluene to hydrolyze sterically hindered esters (Scheme 5.1).

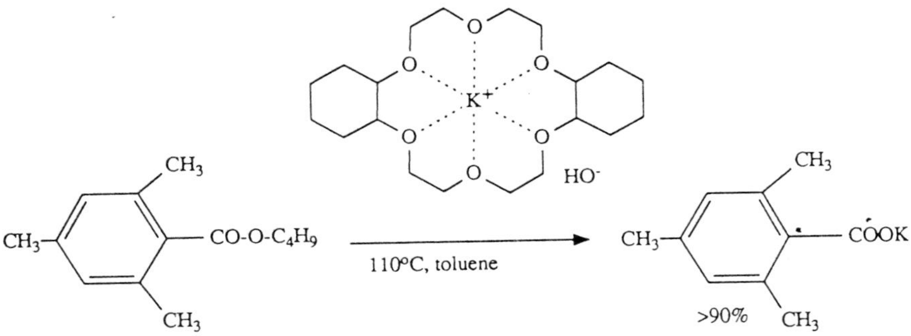

Scheme 5.1

It should be noted that A.C.S. reagent grade potassium hydroxide KOH is 85% KOH and 15% water. On a molar basis, this is almost a one-to-one mixture of potassium hydroxide and water. Thus, the dissolution of potassium hydroxide in toluene probably occurs with a concomitant dissolution of water. This is important because some salts will not dissolve in nonpolar solution in the absence of some water.

Potassium permanganate, $KMnO_4$, is a powerful oxidizing agent but its full utility has sometimes eluded organic chemists because of its solubility properties. It is a highly crystalline purple compound that fails to dissolve in anhydrous benzene. Harold Freedman of the Dow Chemical Company showed that $KMnO_4$ also failed to dissolve in benzene when dicyclohexano-18-crown-6 (**2**) was added. The deeply purple crystals were visible at the bottom of the reaction vessel but no color developed in the benzene solution. It appears that the lattice energy of dry potassium permanganate is too high to be compensated by the solvation energy of the crown. Freedman then added a drop of water to the medium. As he described the experiment to the Author, the reaction was so rapid that color was visible along the pathway of water streaming through the solvent.

This experiment (Scheme 5.2) demonstrates that some source of solvation energy in addition to that provided by the crown is required for complexation to occur. An alternate view is that the water adheres to the salt's surface forming a metastable phase from which ion pair partitioning into the non-polar solvent is facilitated. It seems likely in any event

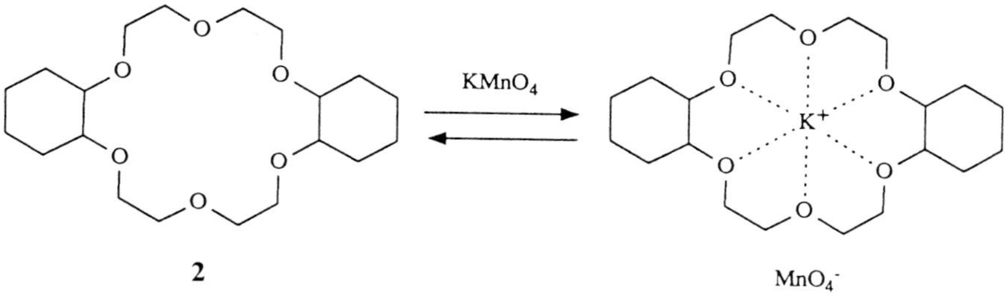

Scheme 5.2

that an interfacial region of some type does exist although at the present time these remain poorly characterized. Of course, water could be involved in both phenomena.

5.2.2 Crown Activated Reagents

Potassium permanganate, when used with crown ethers, is indeed a powerful oxidant. Simmons showed that potassium permanganate in the presence of β-pinene cleanly cleaved the double bond and converted the terpene into *cis*-pinonic acid in 90% yield (Scheme 5.3).

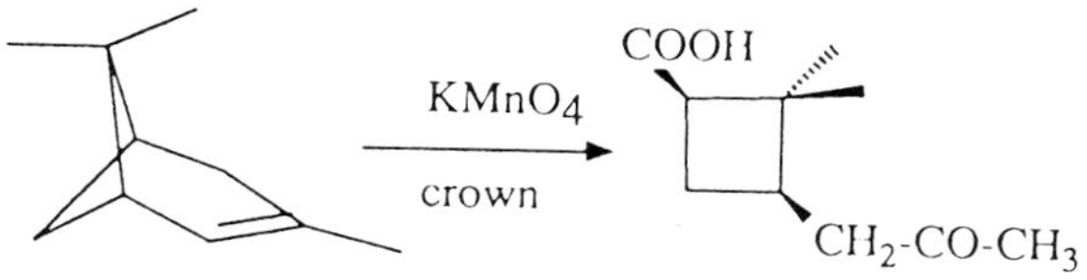

Scheme 5.3

We noted in Chapter 1 that Simmons (Dupont) was working along lines somewhat similar to those that led Lehn to the cryptands. He also recognized the great potential for crown ethers in synthetic applications and realized, for example, that nucleophilic substitution reactions might be effected on systems which had previously proved resistant. Simmons and an associate, Don Sam, heated 1,2-dichlorobenzene (**3**) and potassium methoxide at 90 °C in the absence of solvent (Scheme 5.4). The reaction yielded 2-chloroanisole (**4**) but none of the meta-isomer; this suggested that a nucleophilic aromatic substitution had occurred but not by the benzyne (elimination–addition) mechanism normally invoked for such processes. An equimolar amount of the *meta*-isomer would have suggested a benzene mechanism.

The ability of crown ethers to enhance the rate or alter the course of a given reaction is both interesting and important. In early work conducted from the Author's laboratory, some in collaboration with Bartsch, we

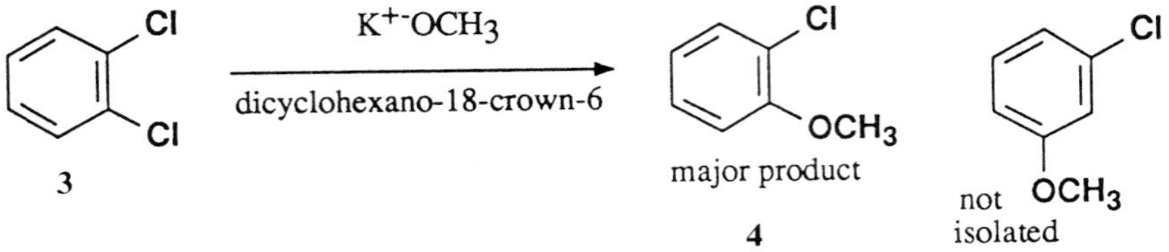

$$\text{3 (1,2-dichlorobenzene)} \xrightarrow[\text{dicyclohexano-18-crown-6}]{K^{+-}OCH_3} \text{4 (2-chloroanisole, major product)} \quad \text{not isolated: 3-chloroanisole}$$

Scheme 5.4

demonstrated that some crown ethers could function as self-solvating bases. The concept of self-solvating bases was originated by Ugelstad in the 1960s. He recognized, for example, that polyethylene glycol derivatives could be deprotonated to form an alkoxide which could then 'wrap itself up.' The result was a salt in which the cation was solvated intramolecularly (Scheme 5.5).

$$CH_3(OCH_2OCH_2)_nOH \xrightarrow{M^+base^-} CH_3(OCH_2CH_2)_nO^-M^+$$

Ugelstad's self-solvating base

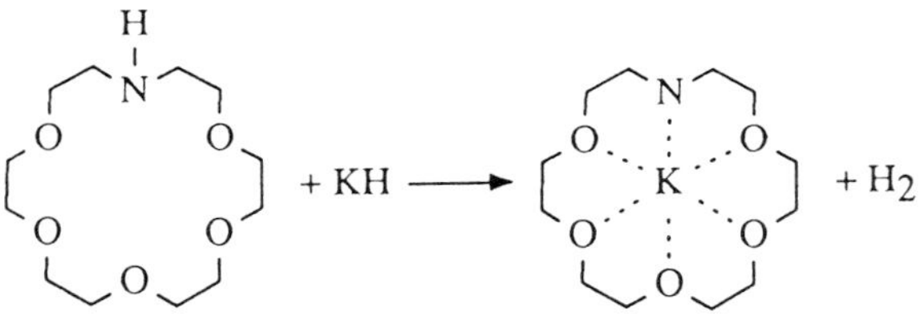

$$\text{aza-18-crown-6 (N–H)} + KH \longrightarrow \text{K}^+\text{-complexed aza-18-crown-6 amide} + H_2$$

Self-solvating crown ether base

Scheme 5.5

'Self-solvation' may not be obvious but it is an important, precursor to the crown ether concept. More important, self-solvation permits molecules to exist as separate ion pairs rather than as part of the aggregates that are common in non-polar solutions. Potassium *tert*-butoxide, for example exists in aggregation states ranging up to about eight units per cluster depending upon the polarity of the solvent. Thus, potassium *tert*-butoxide, in the absence of a crown ether or other solvating agent is a large and bulky complex base. When self-solvated, *tert*-butoxide behaves quite differently (Scheme 5.6).

Bartsch demonstrated the behavior of such materials by examining the elimination characteristics of 2-iodobutane. The results are summarized in Table 5.1. In toluene, potassium *tert*-butoxide is aggregated and therefore reacts as the very large species the aggregate is. Dimethylsulfoxide deaggregates the clusters and this is reflected in both the percent of 1-

Table 5.1 *Elimination reactions of 2-iodobutane*

	% Product DMSO		*% Product Toluene*	
	% 1-butene	*E/Z 2-butene*	*% 1-butene*	*E/Z 2-butene*
t-BuO^-K^+	20	3.5	36	1.7
$Et_3CO^-K^+$	21	3.8	47	1.8
$CH_3(CH_2CH_2O)_7^-K^+$	19	3.4	20	3.7
Aza-18-C-6$^-K^+$	22	3.0	21	3.8

butene formed and in the *E/Z* isomer ratio. The self-solvating systems (last two rows of Table 5.1) are unaffected by solvent showing that they are already 'deaggregated'.

One of the most striking crown ether-induced rate acceleration reactions was reported by Evans and Golob. When the oxy-Cope rearrangement shown in Scheme 5.7 was conducted in the presence of crown ether the rate acceleration was reported to be $\sim 10^5$.

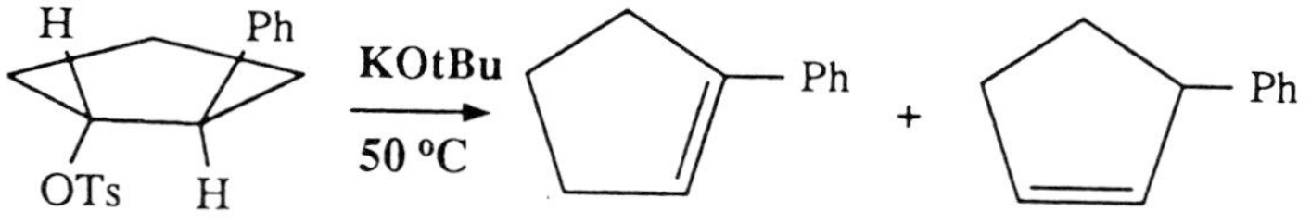

H Ph

OTs H

KOtBu

50 °C

Ph + Ph

product of:	syn-elimination	anti-elimination
no crown present	91%	9%
dicyclohexano-18-crown-6	30%	70%

Scheme 5.6

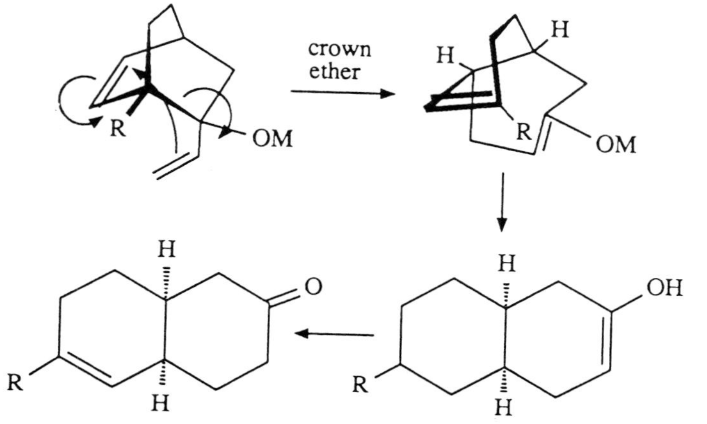

R OM

crown ether

H H R OM

H OH R H

H O R H

Scheme 5.7

5.3 Cation Deactivation

The examples discussed in the previous section demonstrate the remarkable ability of a crown to solvate a cation and thus activate an anion. The complementary process, deactivation of the cation, is also possible although fewer examples are available. If the macrocycle completely surrounds a cation, the latter's Lewis acidic properties will be satisfied by the ligand and will not be available in a reaction. A cryptand should be a better deactivating agent than a crown in this sense because it envelops three-dimensionally whereas a crown does so only in two dimensions.

Furthermore, the equilibrium binding constant, K_S, for the reaction:

$$\text{ligand} + M^+ \rightleftharpoons \text{complex} \qquad (1)$$

is usually greater in the case of a cryptand than for a crown all other variables being equal. The decomplexation rate for cryptands is likewise normally slower than for crowns. A very small reverse rate constant means that even if the equilibrium constant is not prohibitively high, once the cation is complexed, it is essentially trapped. Two examples of cation deactivation are shown here to outline the general principle.

5.3.1 Deactivation of Lithium Aluminum Hydride ($LiAlH_4$)

Some years ago, Pierre and Handel examined the reaction of lithium aluminum hydride with cyclohexanone in tetrahydrofuran solvent (Scheme 5.8). In the absence of a cation-coordinating ligand (other than the solvent), reduction proceeded and cyclohexanol was rapidly formed. When a stoichiometric amount (relative to Li^+) of [2.1.1]-cryptand was added to the reaction mixture no reduction at all was observed. Lithium aluminum hydride is obviously a powerful reducing agent but requires coordination of lithium to the carbonyl in order for hydride delivery to be effective.

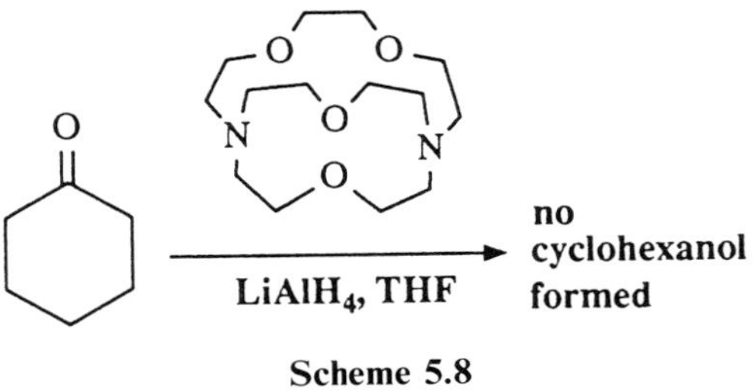

$LiAlH_4$, THF → **no cyclohexanol formed**

Scheme 5.8

5.3.2 Deactivation of the Cannizzarro Reaction

Another reaction that involves hydride transfer is the Cannizzarro reaction. In this reaction, hydroxide ion adds to the benzaldehyde to produce

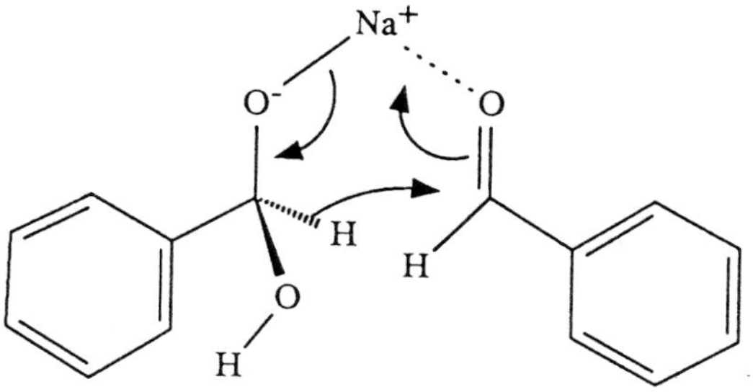

Na+ O- O H H O H

Figure 5.1 *Intermediate in the Cannizzarro reaction*

a geminal diol derivative. This is apparently bridged to a benzaldehyde molecule *via* a metal ion (Figure 5.1). In what appears to be a six electron transition state, there is hydride transferred from the diol derivative to the aldehyde. This results in the formation of benzoic acid and benzyl alcohol. When a crown ether is added to an otherwise successful reaction, the metal ion is competitively complexed and the yield of the Cannizzarro reaction suffers. This is presumably because the crown-bound cation is deactivated and cannot serve as the link to form the ternary complex required for reaction.

5.4 Anion Activation: Phase Transfer Catalysis

The development of phase transfer catalysis parallels the development of crown ether chemistry. Its origins lie a decade earlier in the work of diverse groups. The catalytic potential of such systems was recognized independently by Miecyslaw Makoza in Poland who studied alkylation reactions and by Charles Starks, then at Continental Oil Company in Oklahoma. It was Starks who named the process 'phase transfer catalysis' and who outlined the principles in such a clear fashion that they were almost immediately usable by a broad spectrum of chemists.

At about the same time, Arne Brandström, a Swede, developed a process that he called 'ion-pair extraction' or 'extractive alkylation.' A tetraalkyl quaternary ammonium cation is like a crown or cryptand complex in the sense that the charged species has a large, hydrophobic surface. An anion paired with it in non-polar solution is activated because the anion is poorly solvated either by the cation or by solvent. Brandström used a two-phase mixture such as water and dichloromethane. An inorganic salt such as KCN was dissolved in water and $Bu_4N^+HSO_4^-$ was added to dichloromethane. A metathetical exchange of ions occurred as follows:

$$K^+CN^- + Bu_4N^+HSO_4^- \rightarrow K^+HSO_4^- + Bu_4N^+CN^- \qquad (2)$$

As this equilibrium is established, the soft ion pair ($Bu_4N^+CN^-$) is extracted into the organic (soft) solvent where cyanide anion is poorly solvated and therefore quite reactive.

5.4.1 Principles of Phase Transfer Catalysis

Starks was employed by an oil company at the time he developed phase transfer catalysis. His company was rich in petroleum resources but petroleum is inedible. Starks wondered if the hydrocarbons available in petroleum could be converted into nutritional products. In particular, he wondered if fatty acids could be economically produced from petrochemical derivatives.

As a first step, Starks attempted to prepare cyanooctane. He reasoned that the nitrile residue was a good nucleophile and could readily be used to convert an alkyl halide into the corresponding nitrile derivative. The nitrile could, in turn, be hydrolyzed to a fatty acid which could ultimately be used in foodstuff production. He attempted the reaction without solvent between 1-bromooctane and sodium cyanide:

$$\mathrm{NaCN + R\text{-}Br \rightarrow R\text{-}CN + NaBr} \quad (3)$$

No product was observed after more than a week at reflux temperature. Starks recognized the difficulty: sodium cyanide is not soluble in octyl bromide and *vice versa*. The classical solution to a mutual insolubility problem has been either to use a mixed solvent system (such as alcohol and water) or to use a dipolar aprotic solvent (DMF, DSMO, *etc.*). Neither of these was attractive industrially because both involved additional expense.

Addition of solvent to any chemical process increases the expense of the reaction. First, there is the cost of obtaining the solvent and second, there is the cost in time and energy of removing it to purify the product. Further, addition of any solvent reduces the effective concentration of the two nucleophiles and accordingly the rate of a bimolecular (S_N2) process decreases as the concentration decreases. For the reaction:

$$\mathrm{A + B \rightarrow C} \quad (4)$$

the rate is given by k_2[A][B].

Starks reasoned that he might be able to exchange the poorly soluble sodium cation for a quaternary ammonium or phosphonium cation and thus make more soluble the ion pair. Ultimately, Starks overcame the solubility problem by developing the 'phase transfer catalytic cycle' shown in Figure 5.2.

Sodium cyanide was dissolved in aqueous solution and the alkyl chloride added, although the latter was insoluble in the aqueous phase. When ten mole percent of the tetraalkylammonium chloride was added, an anion exchange process took place that led to formation of tetra-

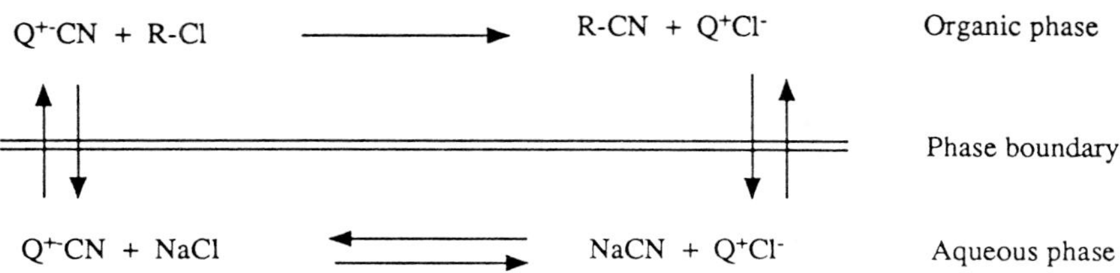

Figure 5.2 *'Phase Transfer Catalytic Cycle' developed by Starks*

alkylammonium cyanide. The latter was soluble in the alkyl halide and reacted rapidly with it. The by-product of the reaction in the organic phase,

$$Q^+CN^- + R\text{-}Cl \rightarrow R\text{-}CN + Q^+Cl^- \qquad (5)$$

is the quaternary ammonium chloride (QCl) (Equation 5) which returns to the aqueous phase and can then exchange with another anion. Thus the process is catalytic.

In most of the work conducted by Makosza, concentrated aqueous sodium hydroxide was used as a base to deprotonate carbon acids such as $C_6H_5\text{–}CH_2\text{–}C{\equiv}N$ to give $C_6H_5\text{–}CH^-\text{–}C{\equiv}N$. This soft anion, which is formed at the phase boundary, pairs with a quaternary ammonium cation (usually $C_6H_5CH_2N^+Et_3$ in Makosza's studies) and then diffuses into the organic phase where alkylation takes place. It was thought for a time that deprotonation or anion exchange reactions occurred in concentrated aqueous solution but this impression was corrected by Montanari and co-workers who showed that such processes occur at or near the phase boundary.

Crown ethers or cryptands function in a fashion similar to quaternary salts. They envelop the cation and make it larger, softer, and more soluble in the organic phase. The phase transfer cycle using a crown can be represented as in Equation 6 in which 'Cr' represents either crown or cryptand.

$$\begin{array}{lcll} [Cr\cdot K]^+CN^- + R\text{-}Cl & \rightarrow & R\text{-}CN + [Cr\cdot K]^+Cl^- & \text{(Organic phase)} \\ \quad\upharpoonleft\downharpoonright & & \qquad\qquad\upharpoonleft\downharpoonright & \\ [Cr\cdot K]^+CN^- + KCl & \rightleftharpoons & KCN + [Cr\cdot K]^+Cl^- & \text{(Aqueous phase)} \end{array} \qquad (6)$$

Two things in particular should be noted in the phase transfer catalytic cycle diagram (Figure 5.2). First, Q^+ represents a quaternary 'onium cation (ammonium, phosphonium, arsonium, *etc.*) but can just as easily represent a crown or cryptand ('Cr' as in Equation 6) complexed metal cation. Thus a few of the large, hydrophobic cations that can serve as phase transfer catalysts are shown (**5–8**).

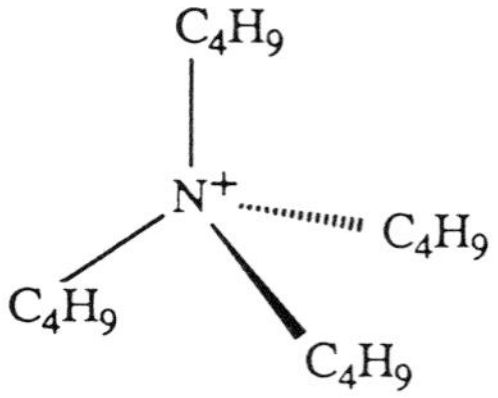

When used in phase transfer processes, the 'onium cation replaces the cation added to the reaction mixture with the nucleophilic salt but the crown or cryptand simply complexes the cation. In either case, a positively charged and hydrophobic species is solvated by a non-polar solution. This cation, Q^+ or $[Cr \cdot M]^+$, provides the anion only weak stabilizing interactions and the anion is therefore more nucleophilic than it would be in water or alcohol.

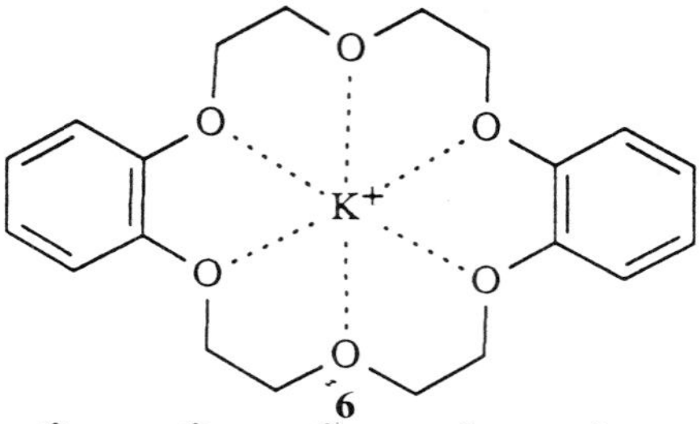

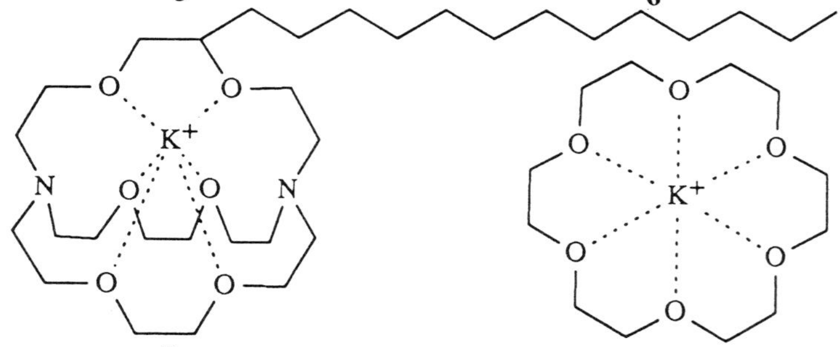

When a quaternary 'onium salt is used as phase transfer catalyst, the anion paired with it is important. The anion should be chosen so that it is both hydrophilic and non-nucleophilic so that it will remain in the aqueous phase and not complicate the substitution reaction. Bisulfate, HSO_4^- is often a good choice. When either a crown or cryptand used as catalyst, no new anion is added so this complication is removed.

One other variable should be mentioned. It is often not so much cation binding affinity that is important in phase transfer catalysis but hydrophobicity (lipophilicity). Thus, dibenzo-18-crown-6 may prove to be a better phase transfer catalyst than 18-crown-6 because it is more hydrophobic even though it is a poorer cation binder for alkali metal ions in homogeneous solution. The extent of complexation is also an important variable, however. The long alkyl chain [2.2.2]-cryptand shown (**7**) was prepared by Montanari and Quici and offers the advantages of high complexation strength and hydrophobicity.

A small number of phase transfer reactions will be noted here as a broad review is beyond the scope of this monograph. The examples are chosen to illustrate special strengths of the method. It should be noted that although each example cited is reported to occur with a crown or cryptand as a catalyst, other catalysts that might work may not have been tried.

5.4.2 Displacement Reactions

Toru and co-workers reported a remarkable displacement reaction of acetate anion. At the time of these studies, the poorly solvated nucleophiles were sometimes called 'naked' or 'bare', although neither designation is really appropriate. Indeed, Makosza suggested that the anions should be called 'bikini' ions since each ion pair was associated with a few molecules of water (*i.e.* it was partly covered), especially when extracted from an aqueous residue. In this case, 'bare acetate' displaces bromide from a cyclohexenone derivative. If elimination of HBr occurred instead of substitution, cyclohexadienone would form and this would immediately rearrange to phenol (Scheme 5.9). Thus the driving force for elimination seems great and yet substitution occurs.

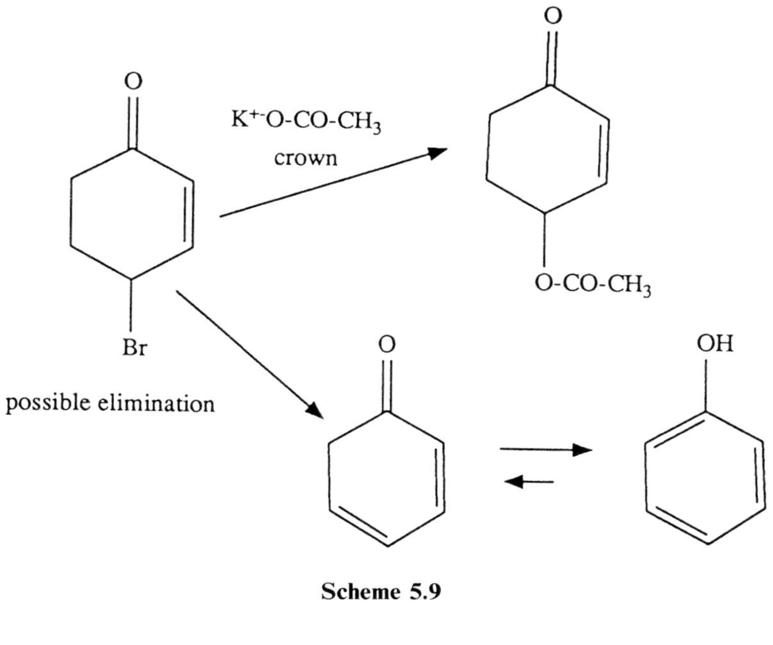

K^{+-}O-CO-CH_3

crown

O

Br

O

O-CO-CH_3

possible elimination

O

OH

Scheme 5.9

The popularity of cyanosilylation [addition of $(CH_3)_3SiCN$ to a carbonyl group] in the early 1970s produced a demand for this intriguing molecule. An early approach to the synthesis of TMS-CN was to treat liquid HCN with butyl lithium and react the LiCN thus formed with Me_3SiCl. The application of a crown ether greatly simplified this procedure and made it much safer (Scheme 5.10).

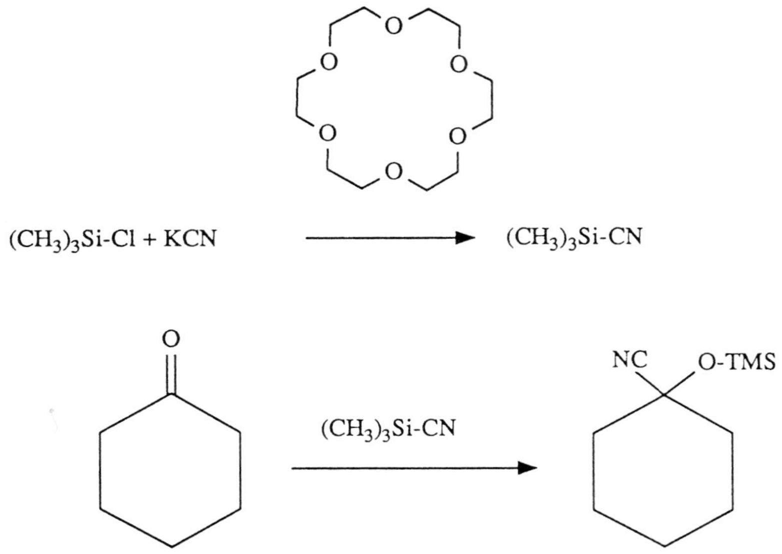

$(CH_3)_3Si\text{-}Cl + KCN \longrightarrow (CH_3)_3Si\text{-}CN$

O

NC O-TMS

$(CH_3)_3Si\text{-}CN$

Scheme 5.10

5.4.3 Superoxide Chemistry

Little was known about the chemistry of superoxide ion (O_2^-) because it is so insoluble in normal organic solvents. Crowns permitted the use of this important reagent as shown in the two examples Schemes 5.11 and 5.12.

In the first case (Scheme 5.11), potassium superoxide affords the product of oxidative dimerization in 52% yield in THF solvent. The second example (Scheme 5.12) is formally a hydrolysis reaction but the nucleophilicity of KO_2 has been used to provide selectivity in this process.

One final example of superoxide chemistry is in order (Scheme 5.13). Rather than being a substitution, it is an addition-induced rearrangement. When tropone was treated with KO_2 and 18-crown-6 in DMSO solution, a 46% yield of salicylaldehyde was isolated.

5.4.4 Aromatic Substitution Reactions

The report by Simmons and Sam (see Section 5.2.2) that 1,2-dichlorobenzene (**3**) underwent substitution with methoxide to give only 2-methoxychlorobenzene (**4**) and none of the *meta*-isomer (expected for a benzyne mechanism) has prompted a number of other studies. Two additional examples are shown in Schemes 5.14 and 5.15. In the second case, two 2-chloropyridine (**9**) molecules were substituted in about 60% yield by 1,6-dihydroxyhexane using KOH as base and 18-crown-6 as catalyst. The

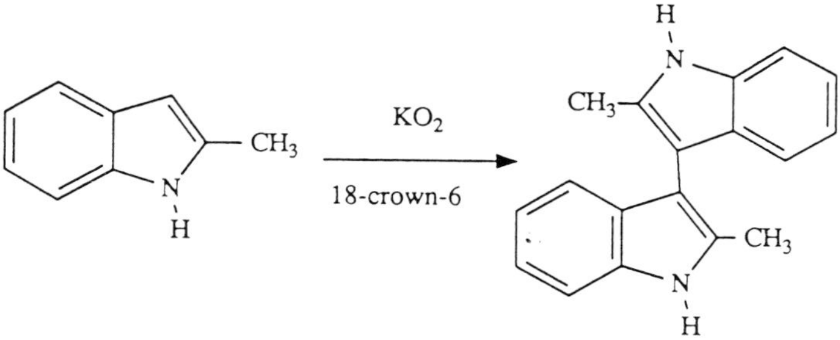
CH3
N
H
KO2
18-crown-6
H
N
CH3
CH3
N
H

Scheme 5.11

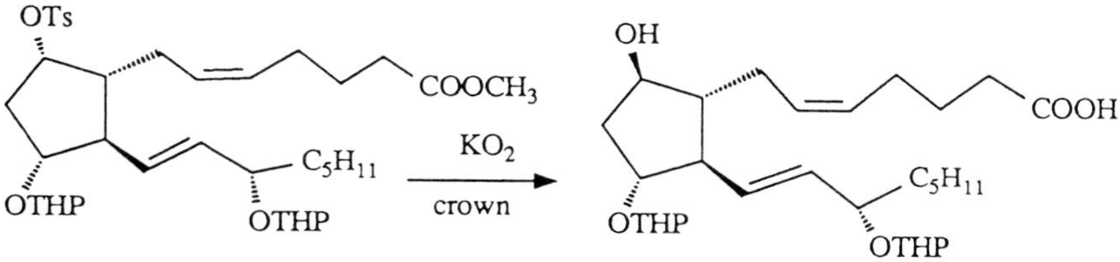
OTs
COOCH3
C5H11
OTHP
OTHP
KO2
crown
OH
COOH
C5H11
OTHP
OTHP

Scheme 5.12

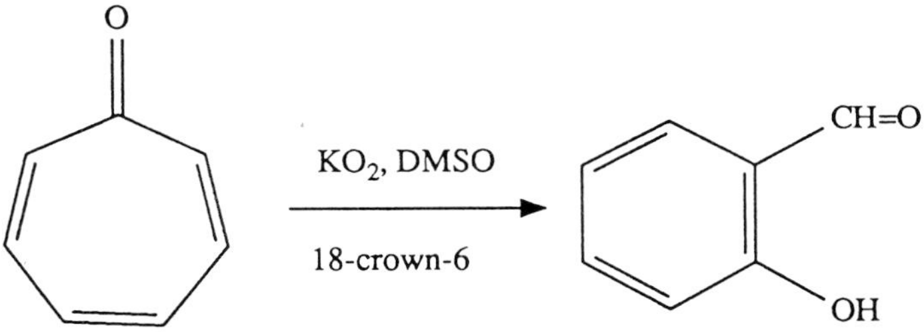
O
KO2, DMSO
18-crown-6
CH=O
OH

Scheme 5.13

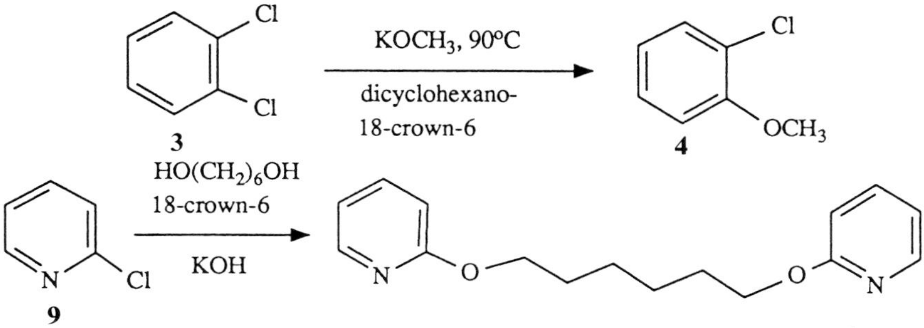
Cl
Cl
3
KOCH3, 90°C
dicyclohexano-
18-crown-6
Cl
OCH3
4
HO(CH2)6OH
18-crown-6
KOH
N
Cl
9
N
O
O
N

Scheme 5.14

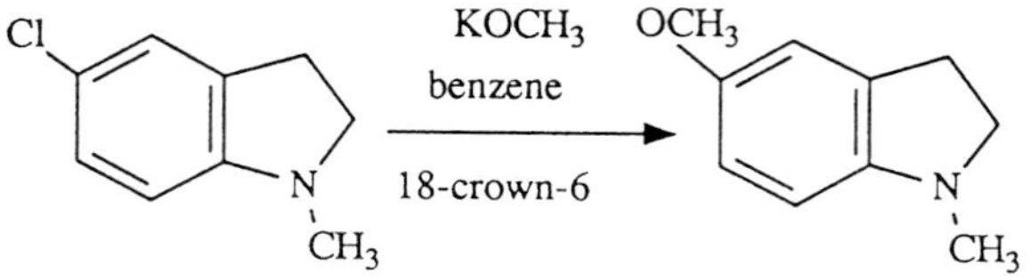

$$\text{5-chloro-1-methylindoline} \xrightarrow[\text{18-crown-6}]{\text{KOCH}_3\text{, benzene}} \text{5-methoxy-1-methylindoline}$$

Scheme 5.15

final case is a more electron rich aromatic that undergoes methoxide substitution in excellent yield (*ca.* 90%) using 18-crown-6 as catalyst.

Nearly innumerable examples of phase transfer catalytic reactions mediated by crowns and cryptands have now appeared. The few examples given are intended only to illustrate the importance of the technique. Monographs and major review articles are cited in the bibliography. Specific reactions for which references are not cited may be found in the Fluka compendia. The interested reader is referred to these sources.

5.4.5 Other Examples

Diazomethane. The formation of diazomethane can be a delicate matter and starting materials are often somewhat expensive. Weber used a clever application of the Hofmann carbylamine reaction to form diazomethane by reaction of hydrazine with dichlorocarbene followed by rearrangement. Hydrazine hydrate could be used since the base was KOH which contains 15% by weight of water anyway. Diazomethane was obtained in nearly 50% yield by this method (Scheme 5.16).

$$H_2N\text{-}NH_2 \xrightarrow[\text{18-crown-6}]{\text{KOH, CHCl}_3} CH_2{=}N^+{=}N^-$$

Scheme 5.16

Use of carbonate bases. Such bases as potassium carbonate are most widely used to scavenge protons rather than as bases to form anions. Crowns can enhance their solubility and their reactivity. A good example is the conversion of phenol into benzyl phenyl ether in quantitative yield by using K_2CO_3 and a catalytic amount of 18-crown-6.

5.4.6 Comparison of Crowns and Quaternary 'Onium Salts

It was once thought that crowns could catalyze phase transfer reactions in which no aqueous solvent phase was used (solid–liquid–PTC). It is now known that quats will often do so as well. Some differences exist, however. Faster reaction rates have sometimes been noted with crowns than with equivalent amounts of quats. The crowns and cryptands are often more stable, especially to base, than are quats but the macrocycles are also much more expensive. It is hard to say at present whether certain reactions mediated by crowns or cryptands are unique. It is certainly true that some reactions are reported to occur only with crowns but whether the corresponding quat-catalyzed reactions would be inferior may be unknown.

5.5 Sensors and Switching

The ability of macrocycles to complex a variety of cations must be considered the key property of crown ethers and cryptands. As discussed previously, numerous methods are available to detect various complexation phenomena. The equilibrium constants for binding have been measured using several different techniques and these are discussed in Chapter 3. An early goal in crown and cryptand chemistry was to see if a macrocycle's inherent cation binding could be altered by a switching mechanism. Another goal that arose very quickly in the macrocycle field was to see if crowns or cryptands could themselves indicate complexation. If so, the crowns could be used to detect or sense cations. If the detection method could be made quantitative, the detection of cations or other species would be facilitated.

5.5.1 Switching Modes

Thus far in the macrocycle field, four general switching modes have been identified. They are as follows:

pH switching (protonation or proton ionization)
Photochemical switching
Thermal switching and
Oxidation–reduction (redox) switching.

Each of these modes has advantages and disadvantages. Each is exemplified in the sections that follow.

5.5.2 Ionization Control or pH-Switching

Ionization control or pH-switching is a means by which a compound's charge state is altered. For the 12-ane-4 compound illustrated in **10**, this would be accomplished by lowering the pH. The four nitrogen atoms in the compounds shown are strong donors, especially for transition metal cations. When the pH is lowered, the nitrogen atoms are successively protonated. This does two things to the system. First, the electron pairs form bonds to hydrogen making them unavailable to bind other cationic species. Second, the nitrogen atom becomes positively charged. The

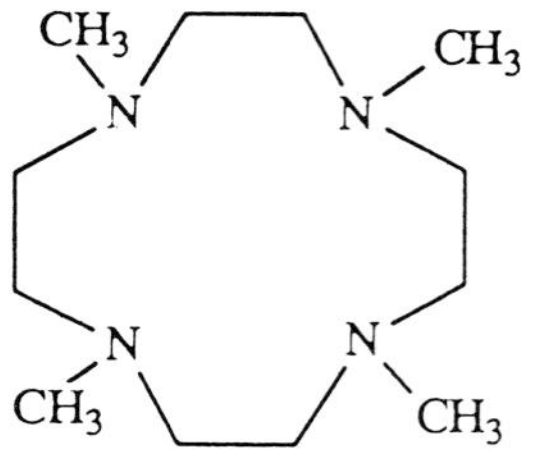

10

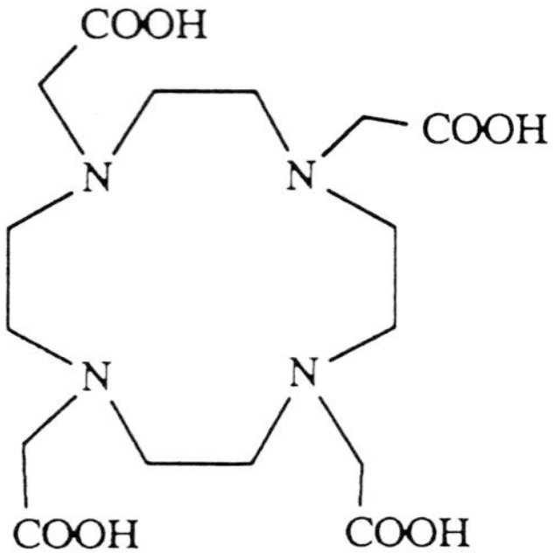

11

charge repels any cationic substance that might be bound by the remaining donor groups even if they are not all protonated. At low pH, then, **10** is a poor ligand but at higher pH it may complex cations.

The compound structure **11** represents a different situation. At neutral or high pH the nitrogen donors are effective for complexing cations but at low pH they are not because tertiary nitrogen becomes quaternary ammonium (subject of course to intramolecular zwitterion formation). When the pH is increased, any protons on ammonium nitrogen are lost again making them available for binding (as with structure **10**). In addition, however, the carboxyl groups ionize to carboxylate anions which are considerably more powerful donors than are the neutral functional groups. Thus the compound **10** is capable of low–high (binding strength) switching as the pH increases from 1–14. The compound **11** is capable of low–high–higher (binding strength) switching under the same circumstances.

The groups of Izatt and Christensen in collaboration with Bradshaw demonstrated two interesting examples of proton-mediated switching. One is a pyridone derivative (**12**) with an ionizable N–H group and the other is calix[8]arene (**13**), in which 1 to 8 phenolic hydroxyl groups may be lost. Both of the compounds were shown to be capable of cation transport through bulk liquid organic membranes (see Section 3.5 for apparatus). In either case, there is a pH differential in the source and receiving phases. When the pH is high, either molecule is ionized and the anions are strong donors. In this form, they may carry cations through the membrane. At the receiving phase, the pH is lower, protonation occurs, and the cation is released.

Lehn and co-workers accomplished the same sort of proton-switched transport by using cryptands in which the neutral species is the strong binder and carrier and the protonated form is poor in both respects. This approach was used, in part, because cryptands are not dynamic binders and do not generally lend themselves to effective cation transport. Scheme 5.17 is not intended to illustrate a particular protonated form of the cryptate complex.

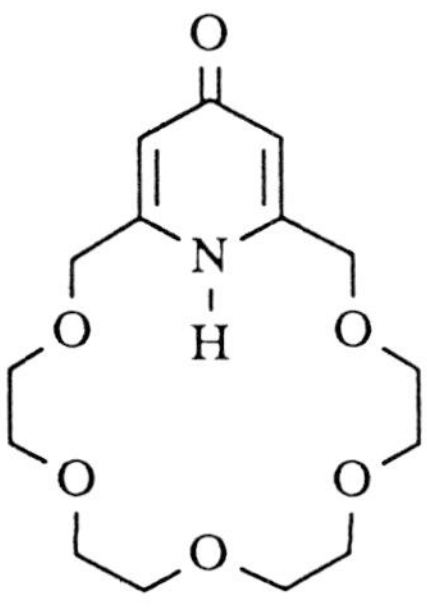

12

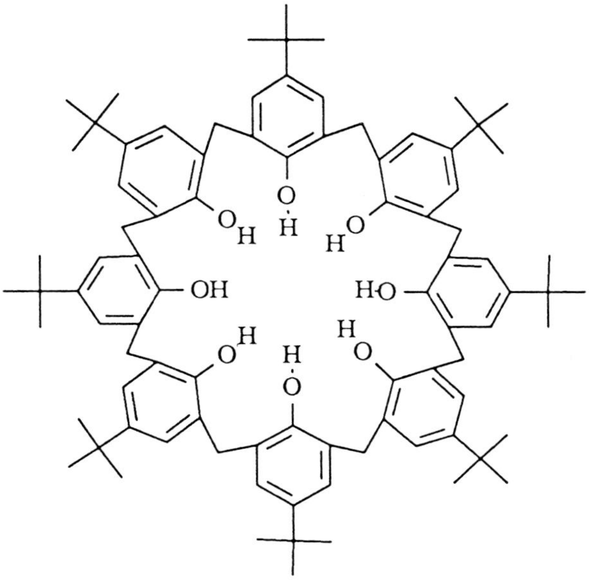

13

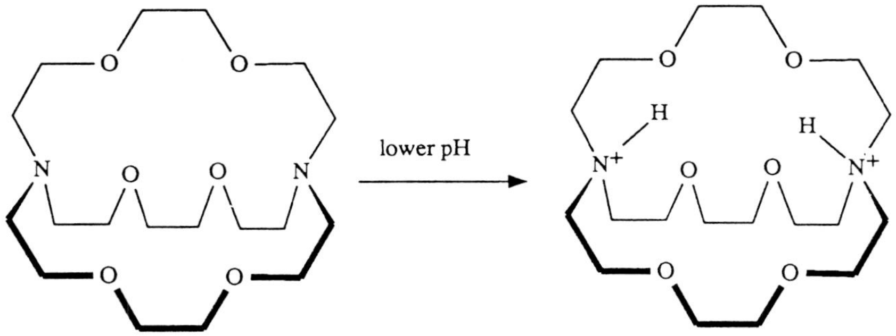

lower pH

Scheme 5.17

An especially important aspect of switching by ionization is the observation that complexation will alter the properties of the donor system. An example of this is found in the ultraviolet spectrum of picric acid which is different from the spectrum of sodium picrate which, in turn, is different from the ultraviolet spectrum of calcium picrate. If a residue such as picric acid can be added to a macrocycle in such a way that an electronic interaction occurs between the chromophore and the ring-bound cation, a spectral change can be observed. Another way to accomplish this is to use a compound having the chromophore on a sidearm. When the cation is bound, the sidearm may interact directly with it to produce a spectral change. Moreover, in the latter case, pH-switching may make the sidearm a stronger donor group.

The nitrophenol, dinitrophenol, and picric acid residues have proved particularly popular in such applications. Numerous compounds having embedded chromophores have been prepared during recent years by the groups of Takagi, Shinkai, Pacey, Misumi, Kaneda, and many others. Two examples are shown in Scheme 5.18 and structure **14**. In the first example, picramine is conjugated to the macroring. When binding occurs to a cation M^+, loss of the relatively acidic proton affords a neutral complex, suitable for transport.

Another example of the same concept is illustrated in **14**. As with the previous system, the complex will be neutral if the proton is removed prior to complexation. The phenolic hydroxyl group is directly within the 18-crown-6 residue. This alters the crown's binding properties and also the effective ring size. The fact that the hydroxyl group is part of the

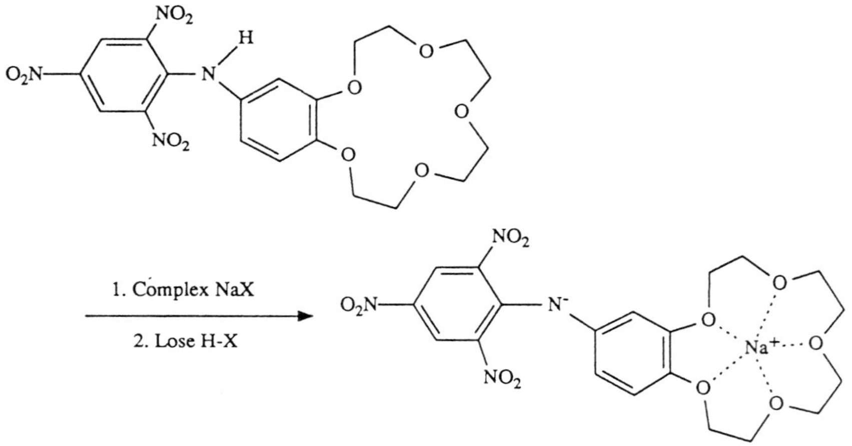

1. Complex NaX
2. Lose H-X

Scheme 5.18

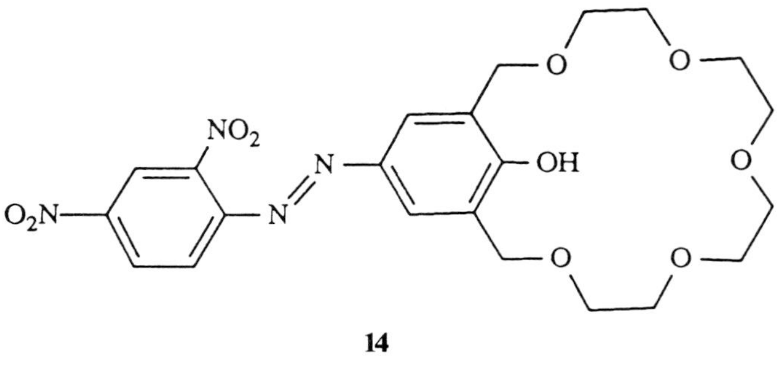

14

chromophore means that a spectral or color change is likely when complexation occurs.

In the next example, Scheme 5.19, the sidearm also provides a charged counter-anion so that the system is again overall neutral and amenable either to colorimetric analysis or transport. In this case, the sidearm is part of a lariat ether and is itself mobile. The macrocycle is shown at left in the neutral, pre-binding arrangement. It is the phenoxide that reaches over and ionizes to provide both binding strength and chromogenic behavior.

For the sake of completeness, it should be noted that several bibracchial (two-armed) crown ether dyes that are diprotonic have been prepared and studied. Particularly interesting results have been obtained for extraction of divalent cations using the *bis*(azobenzene) compound shown in **15**.

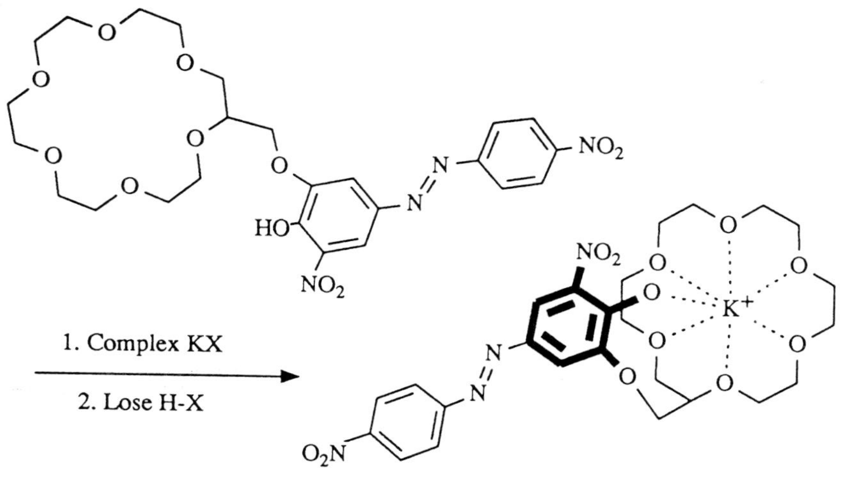

1. Complex KX

2. Lose H-X

Scheme 5.19

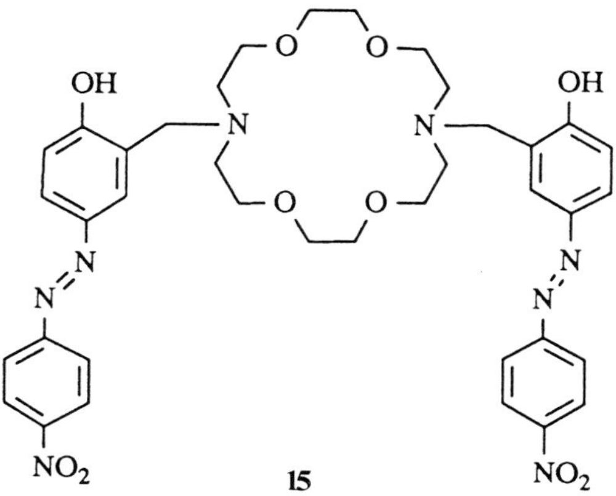

15

5.5.3 Photochemical Control

The photochemical switching mode involves the use of an *antenna* to capture light and to undergo a structural change as a result. The best known example of this is the photoisomerization of *trans* to *cis* azobenzene, Ph–N=N–Ph. The somewhat more difficult photoisomerization of *trans* to *cis* stilbene, Ph–CH=CH–Ph, has also been used but there are fewer examples. Use of the azobenzene residue was exemplified in Scheme 5.19 and **15** where it provided a chromogenic subunit. In this case, both its color (at correspondingly longer wavelength) and its ability to isomerize are of importance. Two examples are shown in Schemes 5.20 and 5.21. The first is a 'back-biting' system developed by Shinkai and co-workers. The second combines the aspects of a photoswitch and a proton ionizable system.

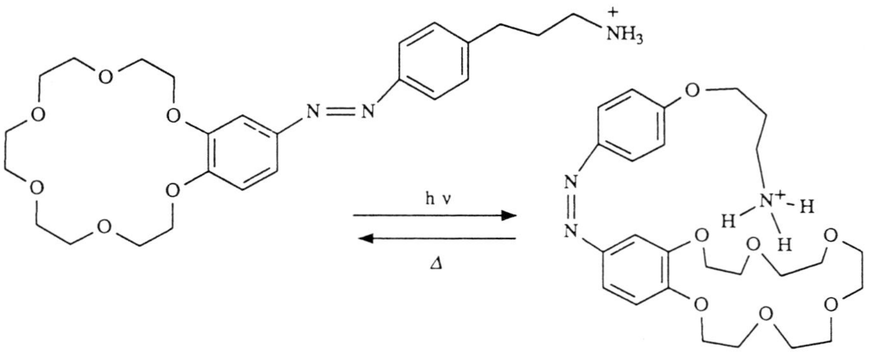

Scheme 5.20

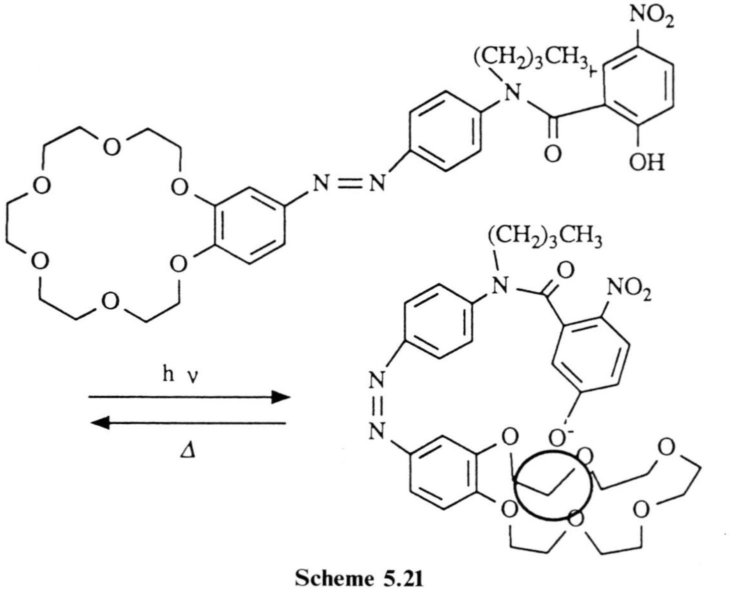

Scheme 5.21

The combination of photocontrol and colorimetric indication of complexation holds forth considerable promise in cation binding and measurement applications. Such sensor or indicator molecules are of especially great potential as simple substrates for assessing ion concentration in human plasma since they require only UV or visible spectrophotometers. Many more examples of these ideas are known and the interested reader should consult one of the reviews cited in the references.

5.5.4 Thermal Switching

The studies reported in this area are relatively limited but potential applications abound. Shinkai and his co-workers devised a membrane system comprising a solid polymeric support, a liquid crystal, and a diaza-18-crown-6 derivative substituted by either a fluorocarbon or hydrocarbon chain. The derivatives containing the fluorocarbon residue form heterogeneous phases within the membrane structure and exhibit higher transport ability when the temperature of the system is raised above the transition temperature for the liquid crystal.

5.5.5 Redox Switching

Three forms of switching have been described: systems whose properties are mediated by changes in pH, by application of light, or by changes in temperature. The fourth type of molecular switching known for macrocycles involves oxidation/reduction (redox) chemistry. Redox switching is based on the principle that a compound existing in one form may be converted by some electron transfer process into another form. For our purposes, these forms will involve states that are altered either to higher or lower binding strengths. Of course, in the relative sense, this may mean changes of the following types: low–high, low–lower, high–low, and high–higher. The designation 'medium' has deliberately been omitted from the discussion since its meaning is even more ambiguous than the other terms.

It is usually the case that addition of electrons to a system (reduction) makes it more electron rich and therefore a better binder for cations. Oxidation, on the other hand, removes electrons with two consequences. The electron richness of the system is reduced *and* a positive charge is introduced. The latter is repulsive to cations and may affect cation binding even more dramatically than the apparent change in electron density.

A redox process may not increase electron density *per se*, but may lead to a chemical reaction that converts a non-binding molecule into a binder. Shinkai demonstrated this principle with the redox-switched *bis*(thiol) $\rightleftharpoons$ disulfide system illustrated in Scheme 5.22.

The examples of compounds in which a neutral system (*e.g.* **16** and **17**) is converted into a positively-charged one are relatively rare. Most of the work done in this area has been reported by Hall, Beer, Kaifer, and Saji

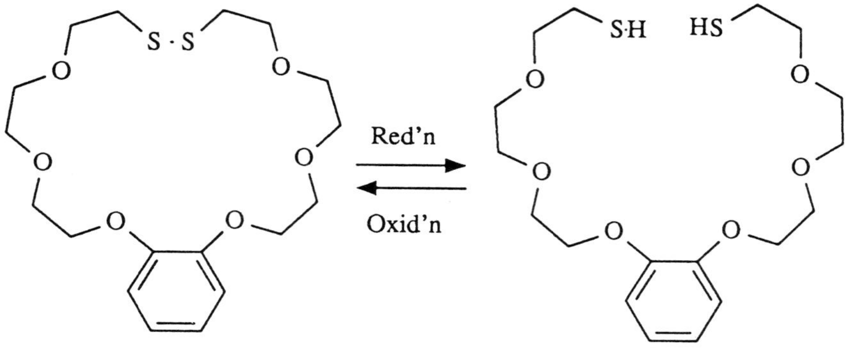

Scheme 5.22

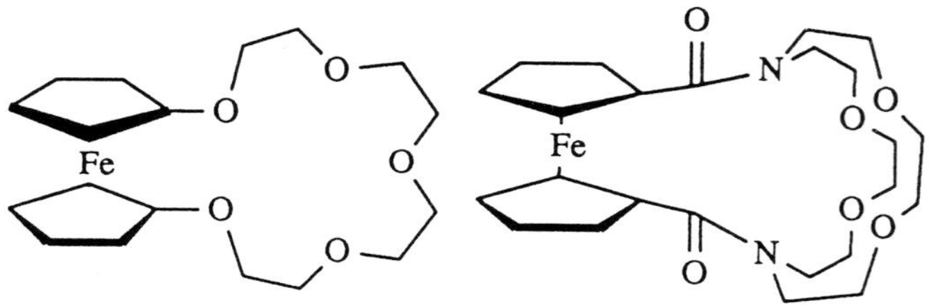

16 17

and their co-workers. Pentaoxa[13]ferrocenophane, **16** has been used by Saji in a transport system in which the neutral molecule is a carrier that is switched to a low-binding state by oxidation of the ferrocene to the ferricinium cation.

The notion of a cation carrier system that could be 'switched on' by electron transfer and ultimately 'switched off' at the other side of a membrane was developed by Echegoyen in collaboration with the Author. Echegoyen recognized that in a compound having a nitroaromatic sidearm, an electron could be added to the aromatic ring and additional electron density would be localized on the nitro group oxygens. With Kaifer and the Author, he showed that when the nitro group was *ortho* (**18**) rather than *para* (**20**), the electron-density-enhanced oxygens would be in a position to bind the cation much more strongly than would be the case with a neutral nitro group (**19**). If the sidearm was in the *para* position it would make little difference whether the nitroaromatic residue was reduced or not because the oxygens of the nitro group could not reach the ring-bound cation (**21**).

The electron transfer to lariat ethers having both *ortho*-nitro and *para*-nitro aromatic sidearms was demonstrated by studies of the electron paramagnetic resonance spectra. Some of the most interesting information, however, was obtained by using cyclic voltammetry. In the cyclic voltammetry experiment, a solution containing a reducible material contacts an electrode, the potential of which is varied. The experiment is described as cyclic because the potential is altered from high to low and then back to the starting point, or *vice versa.*

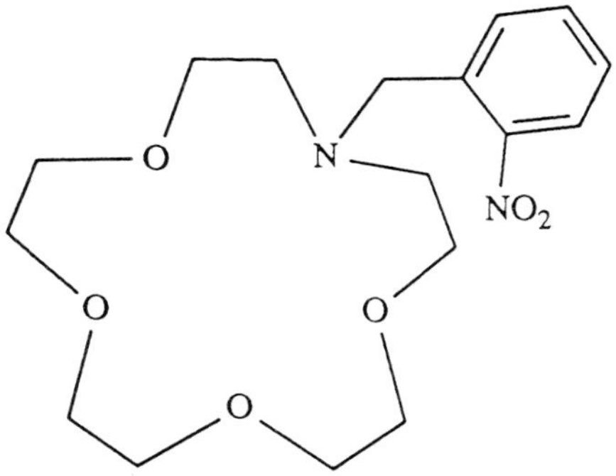

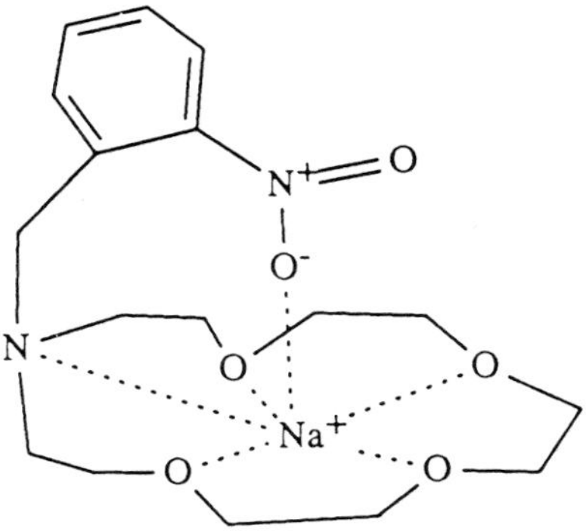

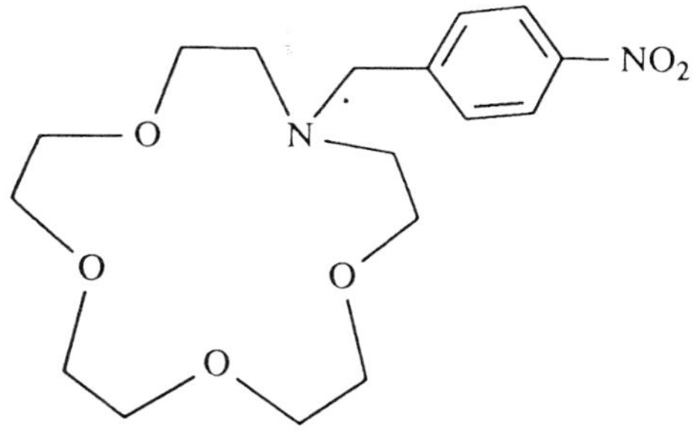

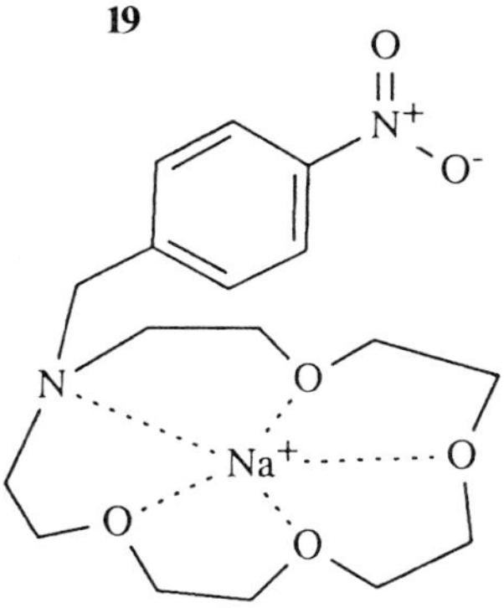

Reversible reduction of the nitroaromatic residue was observed whether the nitro group was in the 2- or 4-position. When a cation was present, however, the situation was different. As shown in the structures **18–21**, the *ortho*-nitro group can interact with a ring-bound cation but the *para*-isomer cannot. In the *ortho* case, the positively-charged, ring-bound cation could exert a remote substitutent effect (*i.e.* act as a Lewis acid) on the nitroaromatic through contact with the nitro group. This was geometrically impossible in the *para* case. Thus, when 0.5 equivalent of cation was present in the system, two redox couples were observed for the *ortho* isomer and only one was observed for the *para* case. When 1.0 equivalent of salt was added, only the new couple was observed for the *ortho*-isomer and no change was observed in the original voltammogram of the *para*-isomer.

These ideas are of use in cation transport across membranes. There is an interesting paradox associated with transport which is due to the simple fact that every membrane is two-sided. For transport of a cation to occur across a membrane boundary, three conditions must apply. First, binding must be rapid and strong at the source phase. Second, the cation must be strongly bound within the membrane. Third, the binding at the receiving phase must be fast but *weak*. A diagram of the membrane system is shown in Scheme 5.23 and these ideas have also been touched on in Chapter 3.

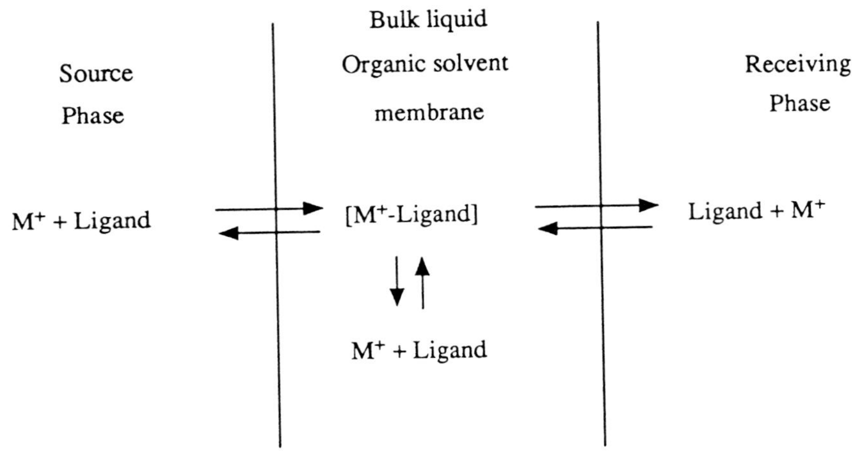

Scheme 5.23

Since it is obviously impossible for a compound to simultaneously exhibit strong binding and weak binding properties, switching of one sort or another has been used to enhance transport. Lehn used a high pH at the receiving phase side to force release of the cryptand-bound cation. Izatt, Takagi, Shinkai, and others have used proton switching in the same way. Electrochemically-switched transport is possible using these same concepts. On one surface, the system is reduced to make the ligand a stronger binder. The strong binder persists during transport. At the opposite membrane surface, electrochemical switching takes place in the form of oxidation and returns the system to its previous low binding state. The intrinsic cation binding ability of the neutral ligand has not been changed by these reactions but the properties of the switched systems differ enough for transport to be enhanced.

A number of structures were studied in order to understand the limits of the electrochemical switching notions. Some of the compounds are illustrated in **22–26**. Often, *para*-isomers that were expected not to be efficacious were also studied to be certain that this assumption was true. The azocryptand (**27**) was originally reported by Shinkai as a photoswitchable system but was found in the Author's group to have excellent properties as an electrochemically-active ligand as well.

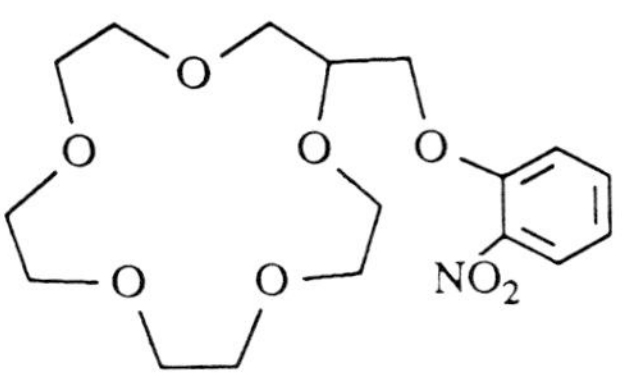

A problem with using nitroaromatics in transport studies is their instability to water. To circumvent this problem, anthraquinone was substituted for the nitroaromatic residue (**28** and **29**). Of course, other changes were made as well because the two are not geometrically equivalent. Further, anthraquinone has the ability to undergo both one- and two-electron redox chemistry. The key to the success of this approach, however, was that the anthraquinone radical anion is stable in water for months so long as O_2 is excluded. The greater binding ability of the anthraquinone system also permitted the use of podands rather than macrocycles, an approach that had failed for the nitroaromatic case (**30–32**).

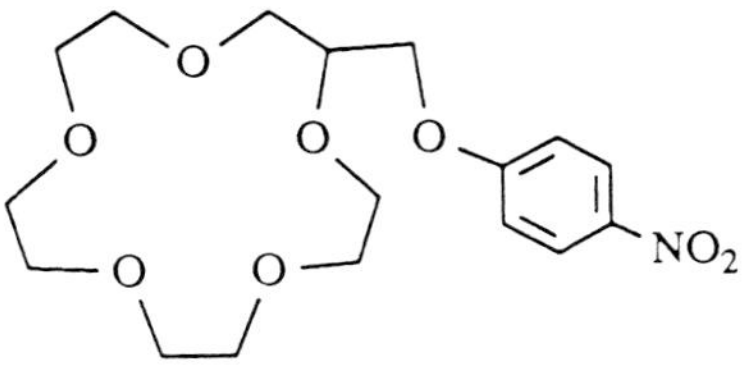

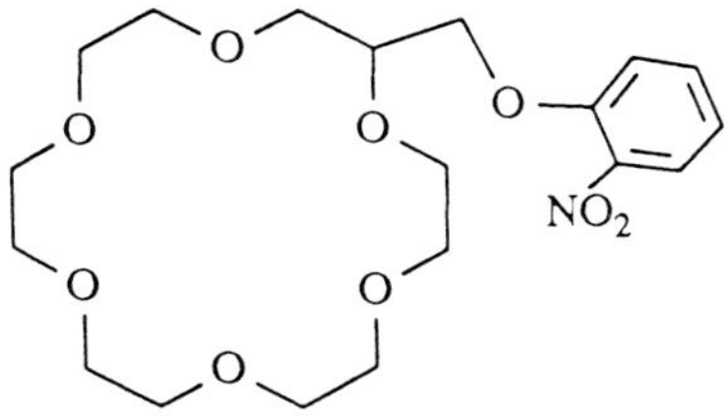

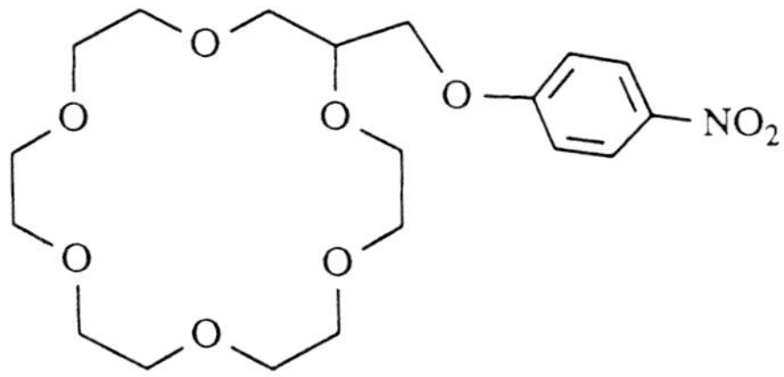

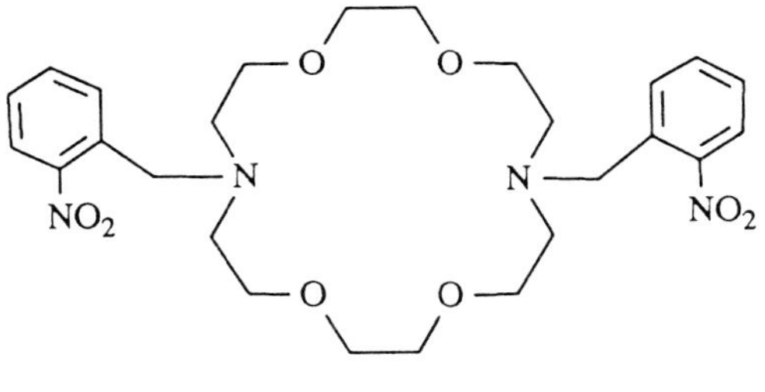

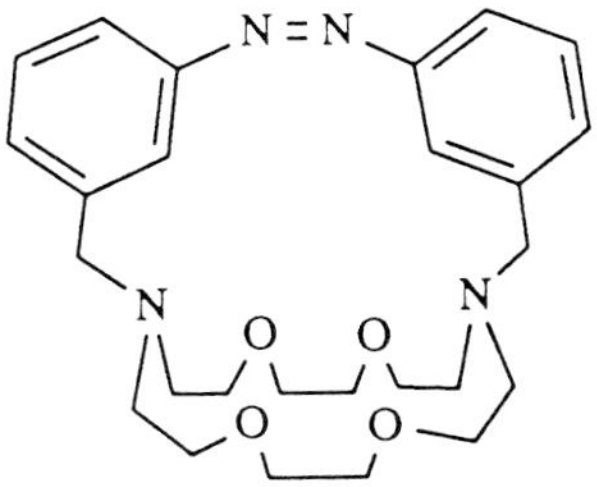

Studies of other radical anion crown ethers (**33** and **34**) were undertaken by the groups of Cooper (**34**) and Bock in collaboration with Vögtle. To the Author's knowledge, these compounds have not been applied in transport studies.

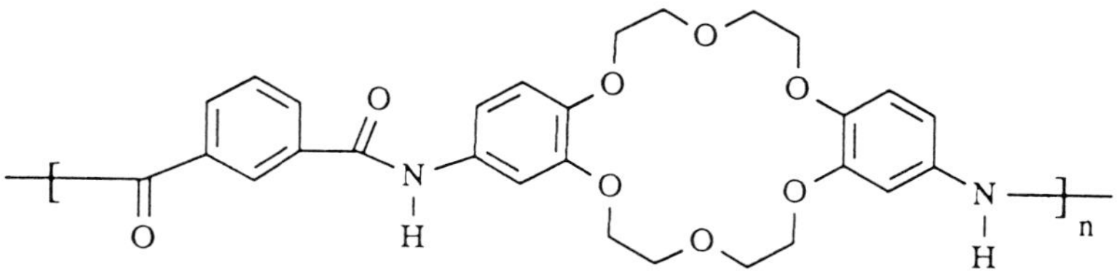

35

R = $(\text{-CH-CH}_2\text{-})_n$ or

$(\text{-CH}_2\text{-CH-})_n\text{COO}$

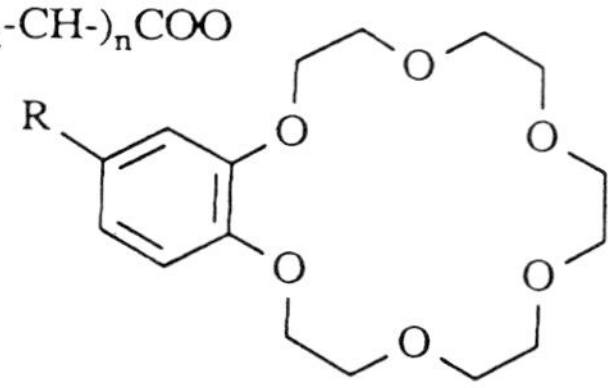

from vinyl or acrylate ester derivatives

36

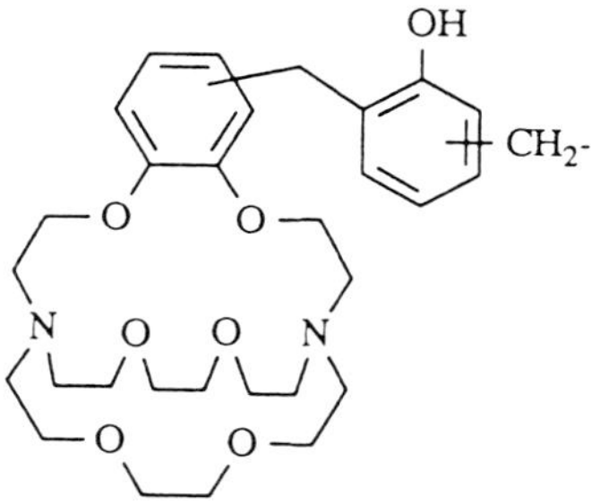

37

crowns. As in peptide synthesis, the spacer chains permit the crowns to extend into solvent while the insoluble beads serve as a mechanical support.

Finally, it should be noted that Blasius and co-workers have used polymeric crowns as analytical reagents. For example, poly(dibenzo-24-crown-8) has been used successfully for ion chromatography that has permitted baseline separation of KCl, RbCl, and CsCl. Likewise, the iodides of these same cations could be resolved.

5.7 Membranes and Channels

5.7.1 Cation Transport

The identification of valinomycin by Pressman as a membrane active complexone and the clarification by Simon of the transport mechanism led to extensive studies of crown ethers as cation carriers. As described in Section 3.5, two devices have been used for most of these studies: the U-tube, sometimes called the Pressman cell, and the concentric tube device. The essential feature of the transport experiment can be understood from the U-tube diagram illustrated in Figure 5.3.

The 'membrane' is typically dichloromethane or chloroform and the source and receiving phases are typically aqueous. Other non-polar solvents that are immiscible with water could also be used but the density of chloroform makes it especially suitable for the U-tube device when used with water. The membrane phase is usually stirred, not to increase the transport rate but to ensure good mixing at the phase boundaries. The transport rate will be affected by the tube diameter and the area of each

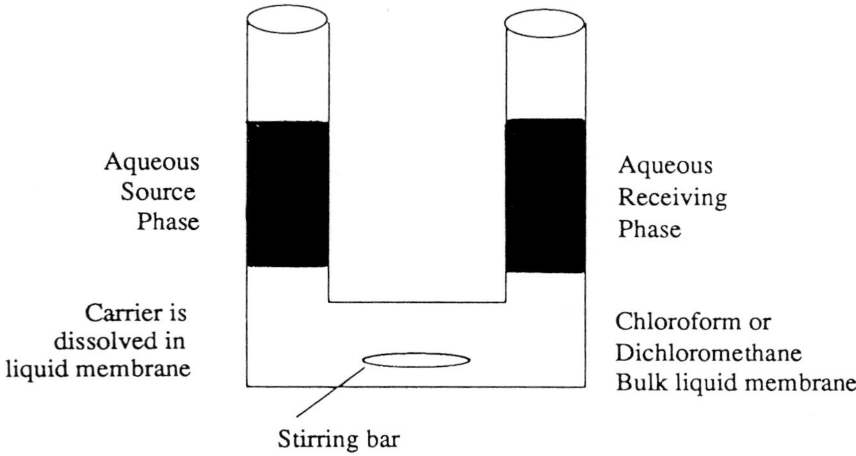

Figure 5.3 *Essential features of the transport experiment to identify valinomycin as a membrane active complexone*

aqueous phase in contact with the membrane phase. Poor stirring or too vigorous stirring (leading to emulsification) will also affect the transport rate.

A crown or cryptand (called the 'ligand' in Scheme 5.2) is added to the membrane phase. Simply put, a salt, MX, present in the source phase forms a crown complex which is soluble in the organic phase. When the complex reaches the receiving phase, the reverse complexation ($[Cr \cdot M]^+X^- \rightarrow Cr + MX$) reaction takes place. The decomplexation and the transport is driven by a solvent gradient. Thus, $[MX]_{source} = \alpha$ and $[MX]_{receiving} = 0$ at the start of the experiment. When $[MX]_{receiving} = \alpha/2$, transport stops because the cation concentration is the same on both sides. By continually removing cation from the receiving phase, *i.e.* replacing it by water, the transport can be pushed further. Each replacement of the receiving phase will permit the source phase to reach a concentration of half its previous value, but as Zeno suggested, it can never reach zero.

The advantage of a switching system (Section 5.5) is thus that transport can be driven beyond the usual 1:1 limit. A difference in pH on the source and receiving sides may not present a switching mechanism *per se*, but provides an antiport to drive the system.

As cation accumulates in the receiving phase or is depleted from the source phase, the extent of these changes can be assessed by using atomic absorption. An even simpler method is to use a colored anion such as picrate for the salt (MX). The appearance of yellow color in the receiving phase is evidence of transport and can be quantitated by colormetric methods.

A vast number of macrocycles has been studied using these devices and related techniques. In many cases, the data are presented as 'per cent of cation transported' in some fixed time (not always specified) and not in terms such as millimoles cm^{-2} s^{-1} or its equivalent that could be compared to the results of other studies. There are many variables such as temperature, relative solvent volumes, *etc.* that also make comparisons difficult. Despite these limitations, much information has been obtained and the interested reader is directed to the reviews listed in the bibliography.

5.7.2 Cation-conducting Channels

Despite the vast number of cation carriers that have been prepared and reported, the preponderent cation transport mechanism in Nature is cation conduction *via* channel formation. The best studied natural system is Gramicidin, a peptide that dimerizes within a membrane and mediates the passage of Na^+. Gramicidin dimerizes and coils into a tunnel shaped channel in which the hydrophobic groups are turned outward and the carbonyl groups are turned inward. The cation, probably in partially hydrated form, moves via carbonyl groups from entrance to exit.

Two basic approaches have been taken in an effort to mimic natural cation channels. These are (1) to mimic shape or structure in the hope that function will also be mimicked, and the other is (2) to attempt functional mimicry directly. The first such attempt was that reported by Tabushi and co-workers in which cyclodextrin molecules were linked together.

Nolte and co-workers found that the isocyanocrown shown in Scheme 5.25 could be polymerized into a cylindrical structure that suggests a channel. Cation conducting properties were not observed for these systems.

Recent efforts have met with somewhat greater success. Notable among these is Fyle's synthesis of a cation-conducting channel based on Fuhrop's bola-amphiphile or bolyte concept. Fuhrhop has, himself,

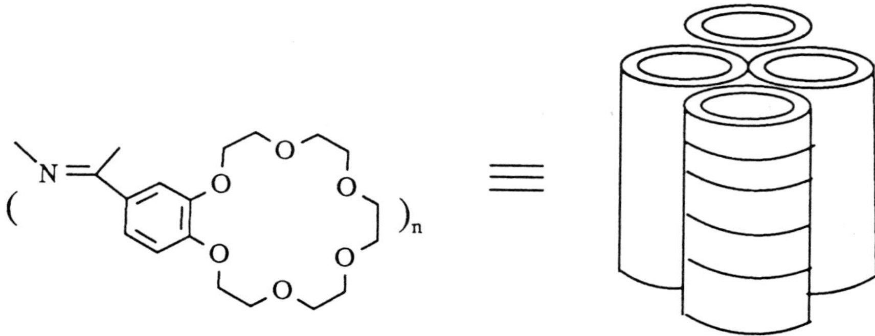

Scheme 5.25

developed an interesting channel although it is beyond the scope of this small volume.

Fyles used 18-crown-6 to serve as the relay in the bilayer midplane. The crown was prepared from three tartaric acid derivatives so that sidearms could be held mechanically in an alternating array. (Earlier efforts from the Lehn group had focused on tartaric acid derived macrocycles as well.) The bolytes, bearing polar head groups thus extended in both directions. Since the bolytes each derive from succinic acid and are twin-stranded, there are six chains extending in each direction from the bilayer midplane. In one case, two octyl chains comprised the bolyte (**38**) and in another, one octyl and one triethylene glycol chain made up this unit (**39**). Both channel-formers were active and the latter was somewhat better, as expected from a cation relay model.

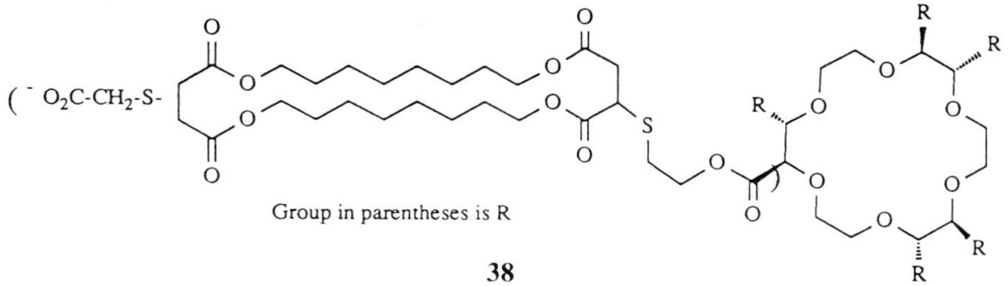

Group in parentheses is R

38

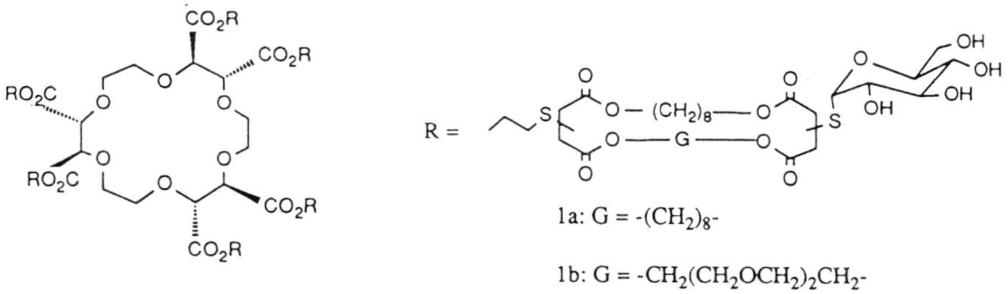

1a: G = $-(CH_2)_8-$

1b: G = $-CH_2(CH_2OCH_2)_2CH_2-$

39

Efforts undertaken in the Author's group also produced a Na^+-conducting channel. The system was designed to be functional rather than to have any structural mimicry of known systems. Indeed, the design was based on the concept that Nature uses the primary amino acid sequence to define a structural framework and then uses conformational effects, van der Waals forces, dipolar interactions, hydrogen bonding, *etc.* to enforce a final, functional structure upon the framework. Thus the present system was designed to have the essential elements thought to be required for a cation-conducting channel. The system consisted of a polar macroring having cation affinity that could lie at each surface of a bilayer. A third ring was placed so that it could reside at the bilayer midplane and serve both as a relay point and a mechanical connection. The sidearms served as mechanical connectors and spacers and also were hoped to provide organization in the way an amphiphile organizes into a membrane.

The structure prepared is shown in Scheme 5.26 in two of many possible conformations. The bottom conformation is 'channel-like' and whether it is realistic or not, the channel proved functional.

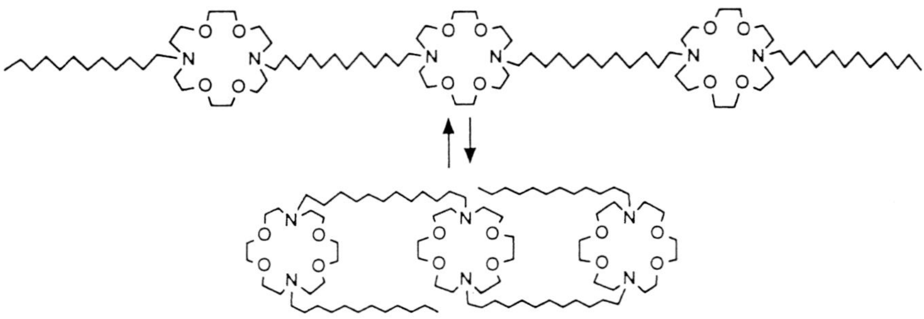

Scheme 5.26

5.7.3 Hydrophobic Macrocycles: Micelles and Membranes

The steroid nucleus is of enormous consequence in natural systems and its stereochemistry is well-defined. It is not surprising that it has intrigued more than one macrocycle chemist. A series of steroidal crown ethers was reported by Stoddart in a review titled 'Chiral Receptor Molecules from Natural Products' (see bibliography for reference). The work was conducted at Sheffield and is reported in the 1979 doctoral dissertation of Pettman. These structures were prepared with an eye to assessing stereochemical consequences in such structures. The cholestan-2β,3β-18-crown-6 derivative is illustrated in structure **40**.

In the Author's laboratory, more or less contemporary with this original design, an effort was made to find crowns having steroidal sidearms that might exhibit liquid–crystalline properties. The notion then was that the rigid and slightly helical steroidal sidearms might organize to give a mesophase. The organization of the sidearms would necessarily cause the macrocycles to orient in a regular fashion as well, and this might produce a cation-conducting channel. The latter goal was not realized in this form, but the steroidal lariat ether compounds did show interesting properties. The compound illustrated in **41** forms vesicles by virtue of the polar macroring head group and the nonpolar steroidal tail. Similar self-assembly (micelle formation) was observed for the amphiphilic cryptands prepared and studied originally by Montanari, Quici, and their co-workers (**42**).

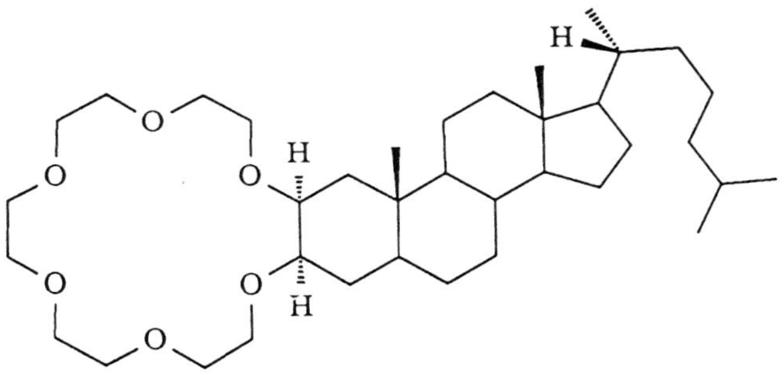

40

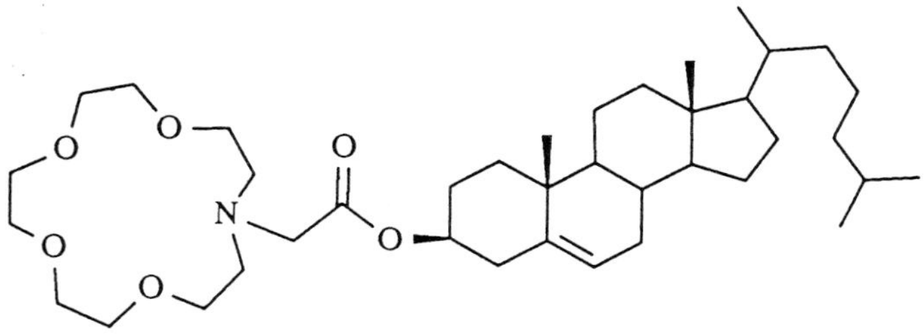

41

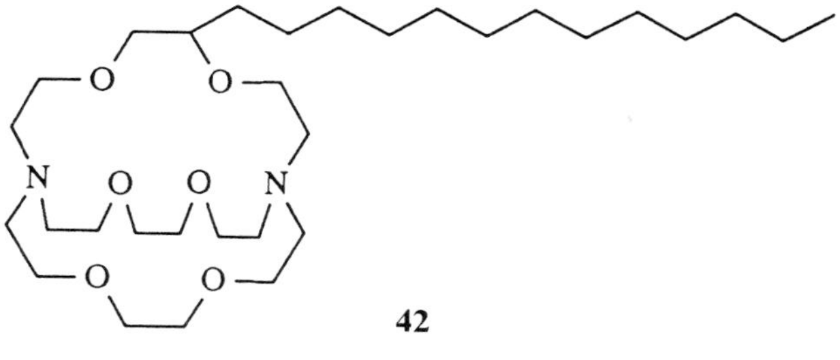

42

The steroidal lariat ethers formed large unilammelar vesicles (LUVs, technically niosomes because the head groups are neutral) by using the detergent dialysis method. These proved essentially similar to those formed by this method from egg lecithin, *i.e.* they were about 3000 Å in size, they were defined by a single bilayer, and they exhibited 1–2% volume entrapments. An important difference was the relative rigidities of these systems. The well-known organizing effect of the steroids made the lariat ether vesicles quite rigid indeed compared to those from egg lecithin.

The cryptands formed micelles rather than vesicles and despite the extremely high cation binding strengths normally observed for cryptands, a large excess (3 to 5-fold) of monovalent cation was required before observable properties suggested that all binding sites were occupied. It is assumed that the cationic cryptand complexes in the micellar assembly repel cations that could fill adjacent binding sites.

Quite recently, Shinkai and co-workers have prepared liquid crystals based on steroidal crown ethers and have shown that addition of alkali metal cations can alter the helical pitch of the mesophase. The structures reported are shown in **43** and **44**.

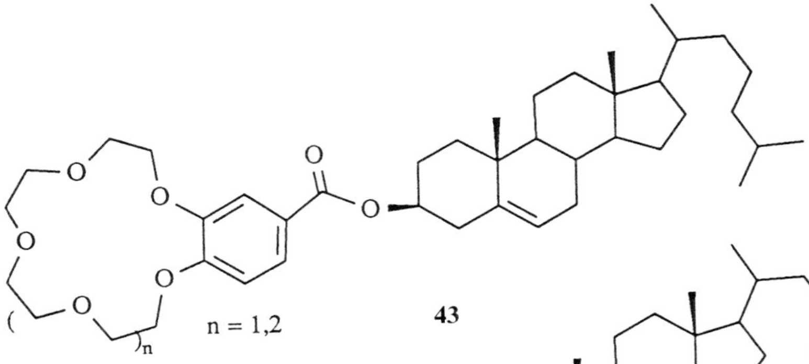

n = 1,2

43

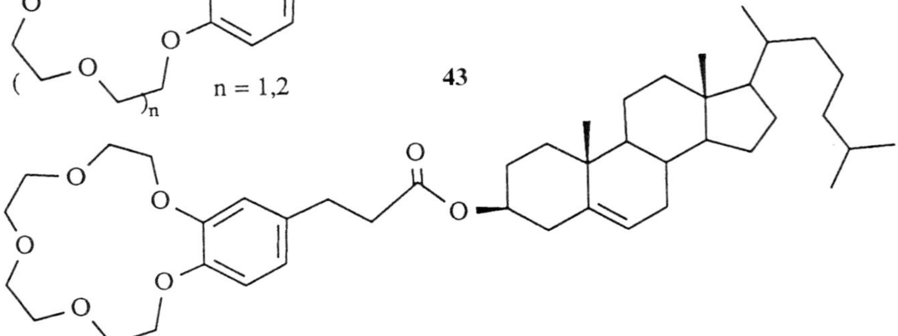

44

5.8 Chirality, Complexation, and Enzyme Models

5.8.1 Cram's Chiral Crown Ethers

Although Cram's dibinaphthyl crown compounds are mentioned in Section 3.6.2, some additional discussion is required to illustrate the importance of the concepts developed in that work. The basic 'chiral barrier' that Cram chose for his studies was the 2,2′-dibinaphthyl system. Ultimately, a number of structures were synthesized, but most were derivatives of the three structures illustrated in **45–47**.

The *tris*(binaphthyl) compound seemed to be the 'ultimate' chiral crown system based on this unit but it proved very difficult to synthesize and problems of rigidity and solubility prevented it from having a more interesting chemistry. The single binaphthyl unit compound was prepared fairly readily and showed some chiral discrimination. Cram had visualized the *bis*(binaphthyl) system as holding great promise because the aromatic units and the macroring were approximately perpendicular. This meant that an ammonium ion could form a tripod of three hydrogen bonds with the ring and three different substituents on an adjacent car-

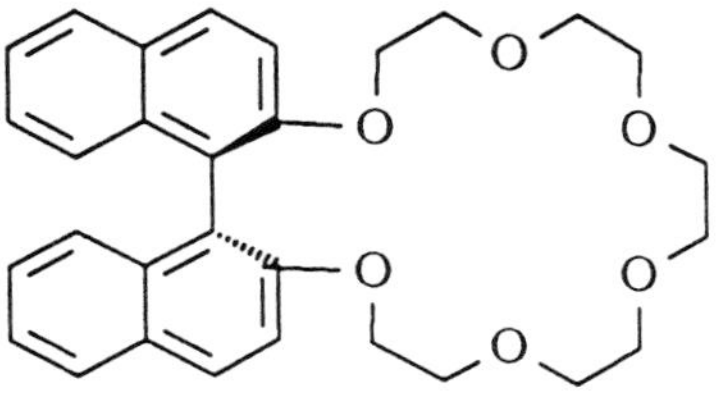

45

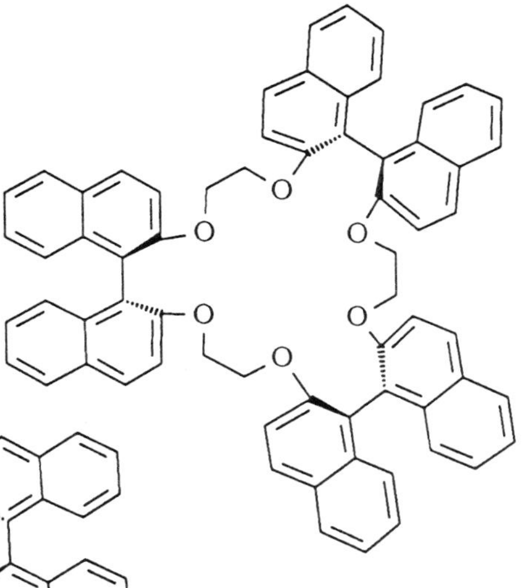

46

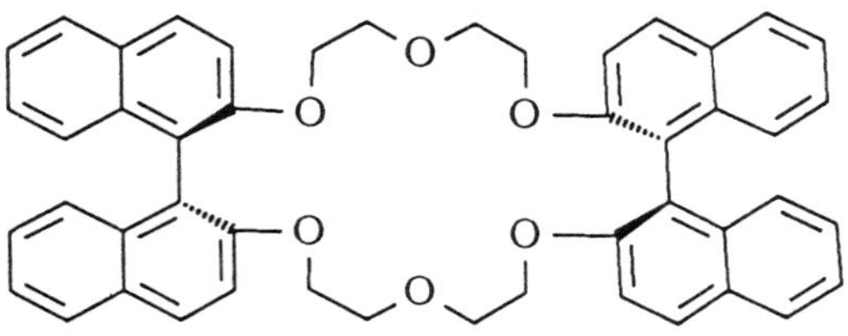

47

bon would find themselves in differently sized cavities. Thus the interaction energies of two enantiomeric ammonium salts with the chiral barriers would afford the complex some selectivity (chiral recognition).

In the semi-schematic illustrations **48** and **49**, all binaphthyl units exhibit S axial chirality. Thus the two crowns are identical and are the *S,S*-isomers. A tripod of hydrogen bonds secures the complexes. In the bottom case (the S amine salt), the large phenyl ring occupies a large empty space while the smaller groups (H and ester) interact with more sterically congested portions of the molecule. In the top case, the large phenyl is forced by the chirality to interact with the bulky binaphthyl unit. Thus the bottom complex is more stable and is favored over the top one. When the chiral macrocycle contacts a solution of racemic amine, a preponderance of the bottom complex forms and this effects a dynamic resolution.

Chiral crown systems designed by Stoddart (**50**), Prelog (**51**), and Lehn (**52**) are shown. The spiro-*bis*-fluorenyl system designed by Prelog is similar to that developed by Cram but is more rigid. In complexation reactions where subtle adjustments of the binding partners are required, rigidity may not always be advantageous. Note also that the *bis*(pyridine) crown reported by Stoddart is based on the tartaric acid unit often used by Lehn.

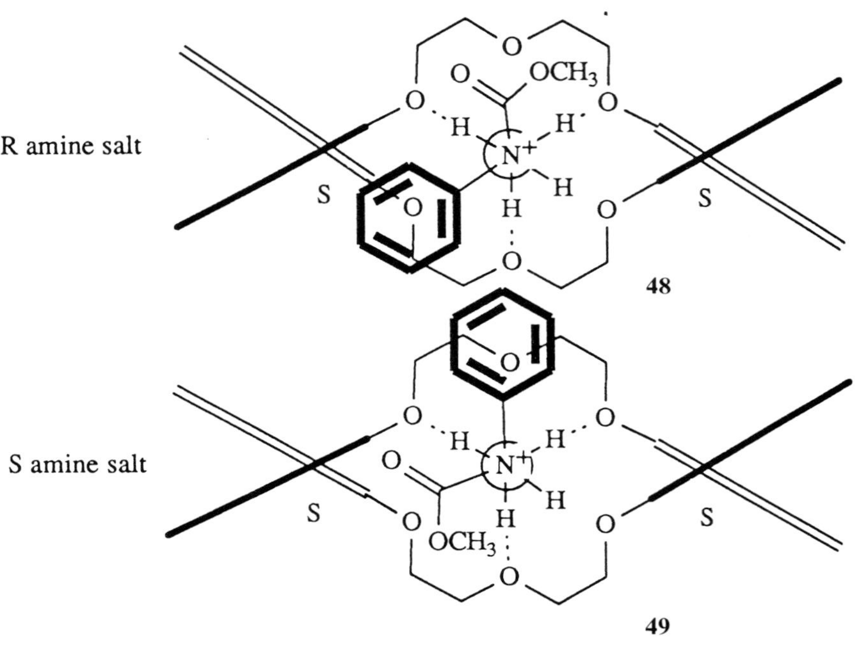
R amine salt
O
OCH3
O
O
O
H
H
N+
H
H
S
O
H
O
S
O
48
S amine salt
O
O
O
H
H
O
N+
H
S
O
H
O
S
OCH3
O
49

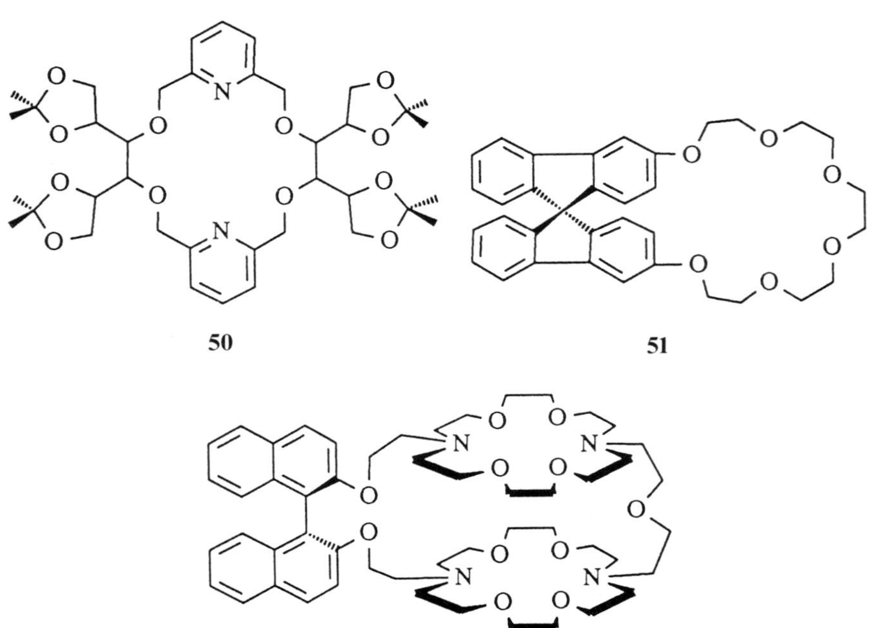
N
N
O
O
O
O
50
O
O
O
O
O
O
51
N
N
O
O
O
O
O
O
N
N
O
O
O
O
52

5.8.2 Rebek's Allosteric System

Rebek and co-workers had demonstrated the presence of allosteric effects in macrocyclic polyethers. The bipyridyl system, *e.g.* (**53**) has two binding sites with different affinities. The 2,2′-bipyridyl unit may complex transition elements and the polyether ring favors alkali metal cations. Complexation at the nitrogens by $PdCl_2$ or $W(CO)_6$ alters the macroring size and overall flexibility. When, for example, $Hg(CF_3)_2$ is bound in the macroring, a rotaxane-type structure is formed. When $PdCl_2$ is bound to the pyridyl nitrogen atoms, dissociation of this complex is reduced by a factor of about 7-fold.

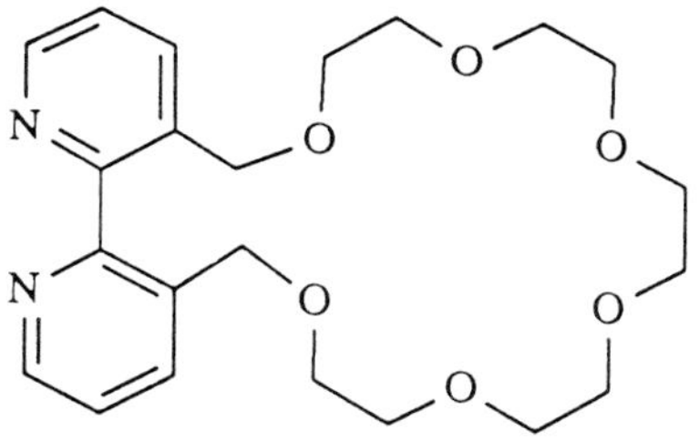

53

5.8.3 Cram's Protease Model

Numerous attempts have been made over many years to understand and then to mimic the activity of the chymotrypsin family of protease molecules. Chymotrypsin's active site contains a complexing cavity, a serine hydroxyl group that serves as a nucleophile, an imidazole, and a carboxyl group (Scheme 5.27). These residues are organized within the enzyme so that they can co-operate in what is often referred to as a 'charge-relay' mechanism. Two decades ago, Bender had attempted to mimic the behavior of these enzymes by using cyclodextrin derivatives. The organic community has been intrigued with these possibilities ever since.

In earlier studies, Cram and co-workers had shown that wonderfully symmetrical, spherical, and rigid host molecules could be built up from phenol subunits. These spherands are all more or less similar to the compound shown in **54** but they may be modified in almost infinite variety.

Cram recognized that in order to mimic chymotrypsin, it would be necessary to enhance the spherand's ability to hydrogen bond an ammonium salt. To do this, he replaced three of the six aromatic methoxy groups with urea oxygen atoms. It has been well-established by the work of Reinhoudt and others in the macrocycle field and Hamilton and others in the 'receptor' field that amides or ureas are superior for hydrogen

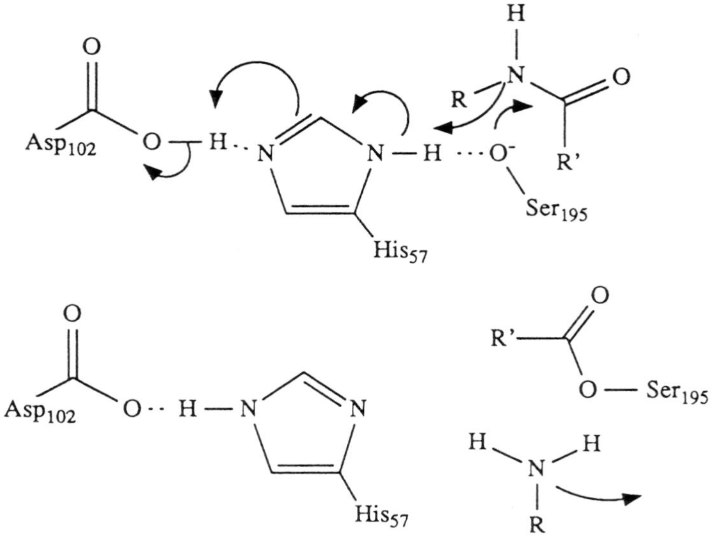

Scheme 5.27

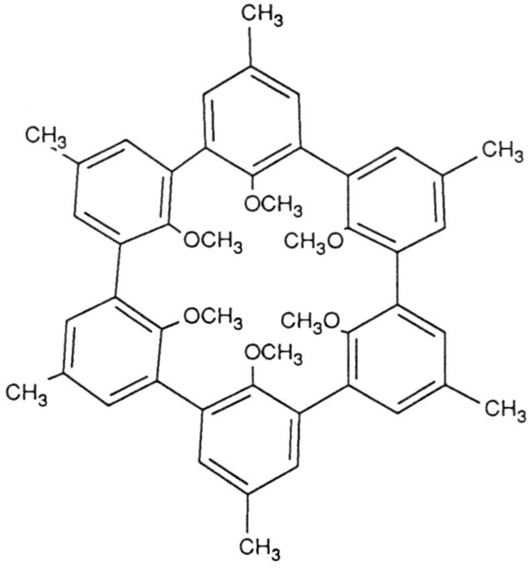

54

bonding interactions. Thus was born the relatively simpler structure shown in **55**. Indeed, it complexed alkylammonium ions very successfully. The proton donor which mimics the serine hydroxyl group was installed on a biphenyl to force it into a proximate geometry **56**. It is interesting that Cram's system has been criticized on the grounds that the functional group geometries are not achieved economically. Be that as it may, the urea residues capture the ammonium salt and the hydroxy group is within distance for an acyl transfer.

Cram demonstrated the efficacy of his synthetic protease by using it to cleave glycine *p*-nitrophenylate perchlorate. In the absence of any added salt, the difference in cleavage rate (determined by release of the colored nitrophenylate anion) between the protease and 3-hydroxymethylbiphenyl was 10^{11}. In the presence of a mole of added $NaClO_4$, the rate diminished slightly to 10^9. These observations are shown in Scheme 5.28.

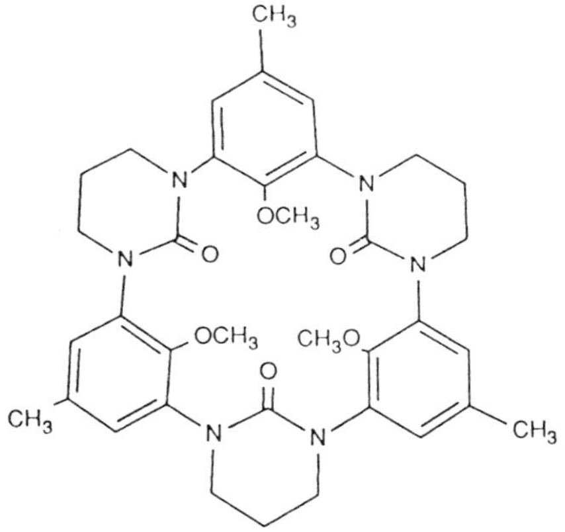

55

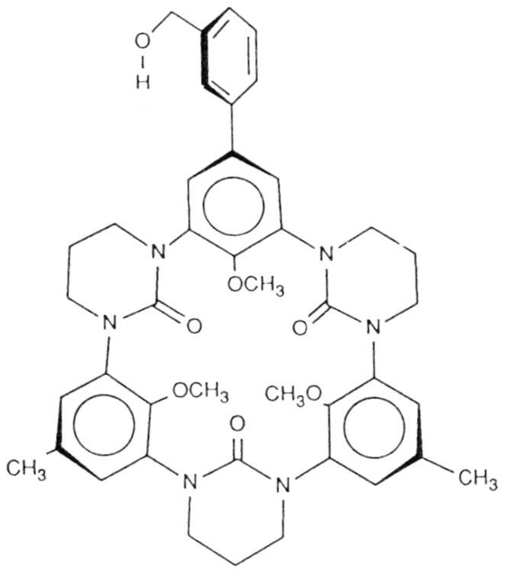

56

5.8.4 Receptor Molecules

Lehn has described the development of cryptands as an effort to effect spherical recognition of cations. This successful effort was described in Chapter 3 and complex structures were illustrated in Chapter 4. Using the cryptands as a starting point, Lehn and his co-workers then proceeded to develop receptors for other geometries. The first such attempt was to prepare a receptor having an internal tetrahedral geometry.

Two interactions were utilized in developing the ammonium ion receptor shown in **57**. Note that the ammonium ion hydrogen atoms are oriented in a tetrahedron and will be directed to the four nitrogen atoms

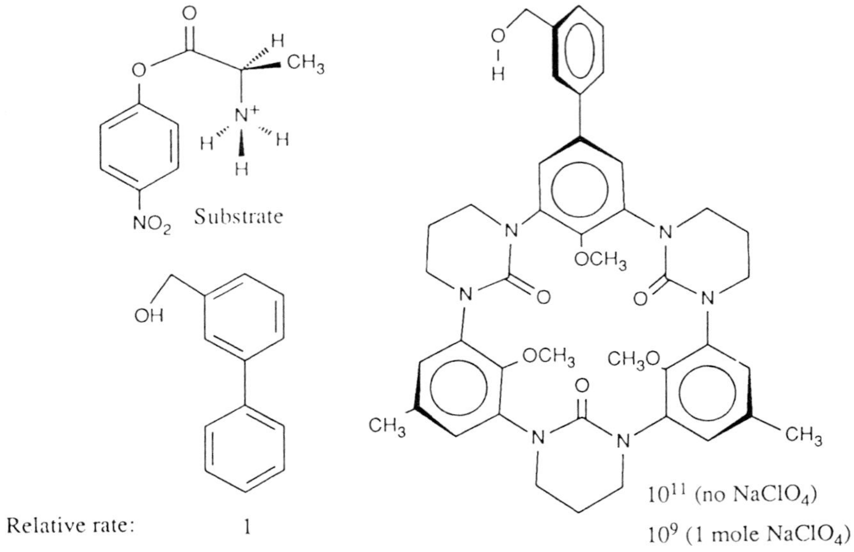

Substrate

Relative rate: 1

10^{11} (no $NaClO_4$)

10^{9} (1 mole $NaClO_4$)

Scheme 5.28

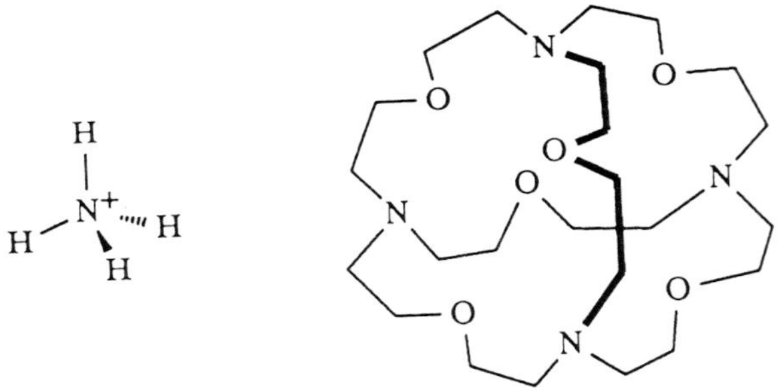

57

when complexed. Three or more of the six oxygen atoms can also approach the positively charged nitrogen atom and provide additional stabilization by a pole–dipole interaction.

Lehn then turned his attention to linear recognition. The *bis*-tren cryptand shown in **58** (see also Section 3.8) can be hexa-protonated to give a receptor of appropriate size and shape to bind azide anion. Likewise, a bromide anion could also be bound in this cavity.

Linear recognition of diammonium ions was accomplished by placing two 18-crown-6 molecular frameworks face-to-face. Rigid spacer units such as 1,4-phenylene, 4,4′-biphenyl, and 4,4′-terphenyl provided spacing. The aromatic residues also exert a magnetic influence on the bound ammonium salt and this was used to advantage by Sutherland who

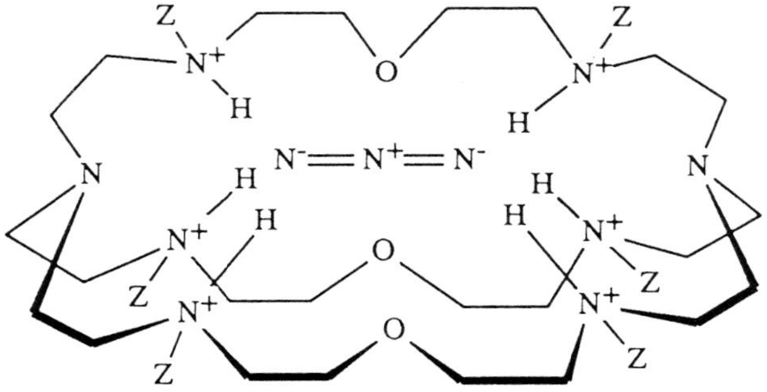

58

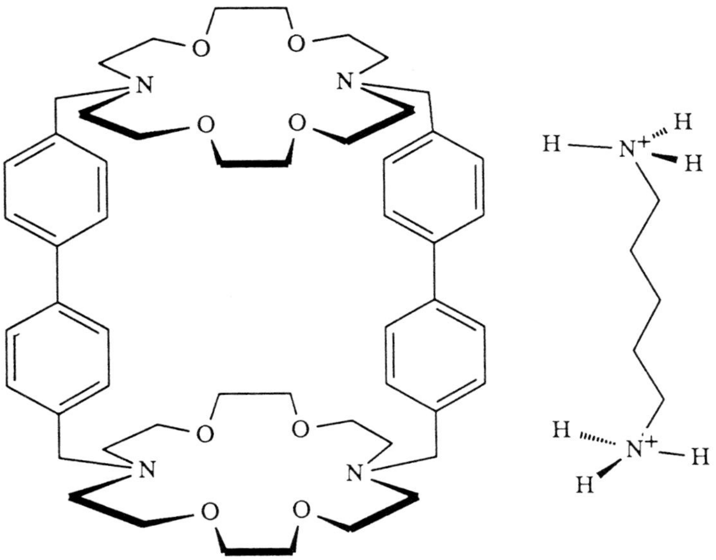

59

assessed binding by detailed NMR studies. In the structure illustrated, **59**, 1,5-pentanediammonium cation is pictured with the receptor that favors it.

When complexation occurs, three hydrogen bonds form between each ammonium ion and each macroring (see Section 3.5.1). Complexation can be detected by the upfield shift of methylene groups held against the aromatic 'sides' of this molecular box. Different aromatic spacer groups give different diammonium ion selectivities and different portions of the aliphatic chain showed upfield shifts depending upon which methylenes are closest to the benzene shielding zone in each particular case.

Schmidtchen has developed an interesting family of receptor molecules in which a tetra-ammonium cation provides an anion binding site and a sidearm provides additional complexation. An example is shown in **60** in which ammonium binding is accommodated by a triaza-18-crown-6 residue and four 8-carbon chains [*i.e.* Z = $(CH_2)_8$] separate the quaternary ammonium salts in the tetrahedral receptor.

It is believed that when complexation occurs, the anion inserts in the positive tetrahedron and three N–H–N hydrogen bonds occur between the ammonium cation of the guest and the triaza-macroring of the host. Studies were conducted in aqueous solution at mildly alkaline pH and binding was assessed by UV when appropriately substituted phenolic residues were used. Association constants for the receptor and substrate shown were in the range 10^2.

Another ditopic receptor having differentiated binding sites was reported by Saigo and co-workers. It is a macrotricyclic receptor in which the binding units are a crown ether and a cyclophane. The former is expected to interact with an ammonium cation as in the example **60**, and the latter provides a generalized hydrophobic interaction. The structure and a complexation partner are pictured in **61**.

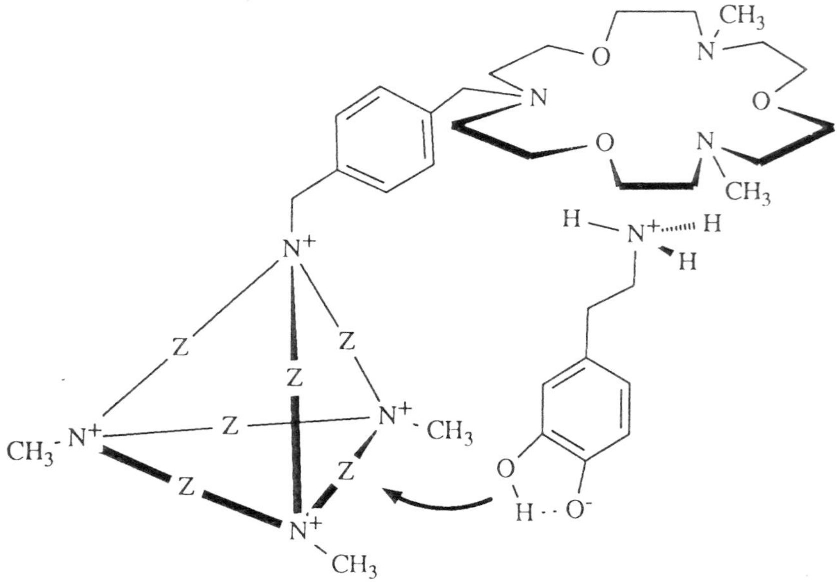

60

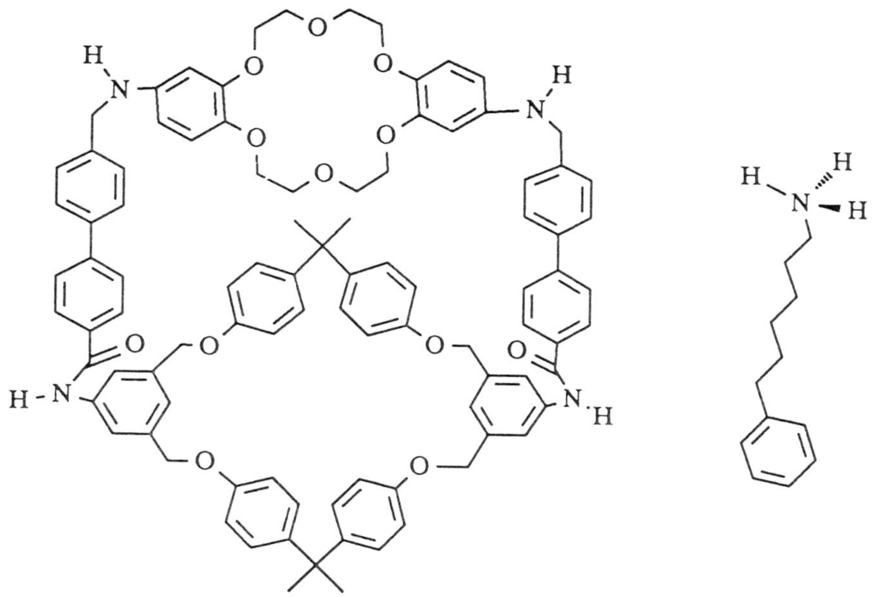

61

5.9 Miscellaneous Applications and Developments

At least since the time of Lüttringhaus, there has been interest in constructing interlocked ring systems. Many attempts have been made and there has been some success. Far more efforts failed than were successful. The three examples shown below are spectacular successes of clever approaches to catenanes.

5.9.1 Metal-templated Catenane Formation

Sauvage, a former co-worker of Lehn and co-author on the original cryptand papers, recognized that copper–phenanthroline complexes were highly organized systems. He thus used them to construct a catenane from the 'inside out'. Once the copper complex was formed, the ends were threaded together (*e.g.* **62**). An even more elaborate if only equally remarkably system was prepared by dimerizing two catenanes using acetylene dimerization (*e.g.* **63**).

Lehn used 2,2′-bipyridyl in a somewhat related approach to make helically coiled copper complexes (Scheme 5.29). While the latter case is not strictly a crown ether, it shows an important future direction of crown, cryptand, and other medium-sized macromolecules. Lehn has called these helix-forming systems 'helicates'.

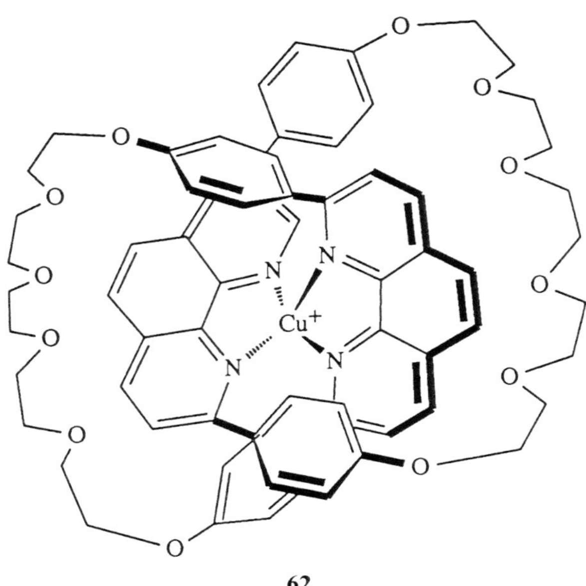

62

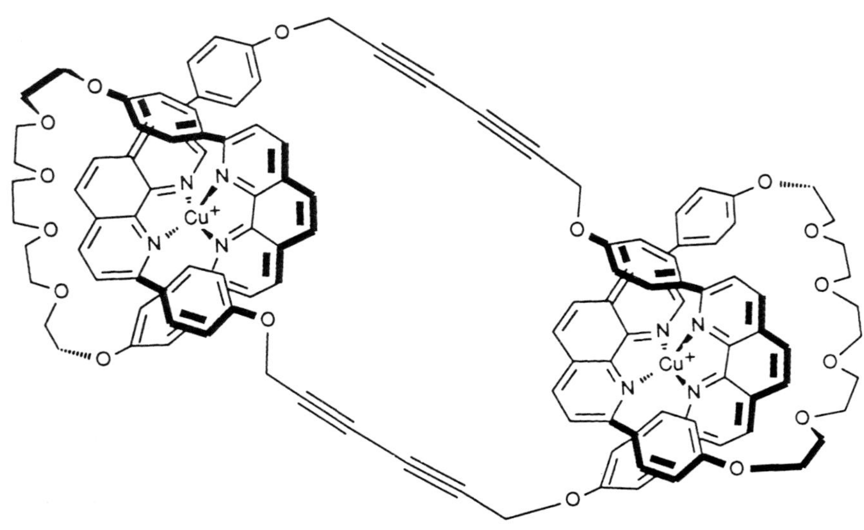

63

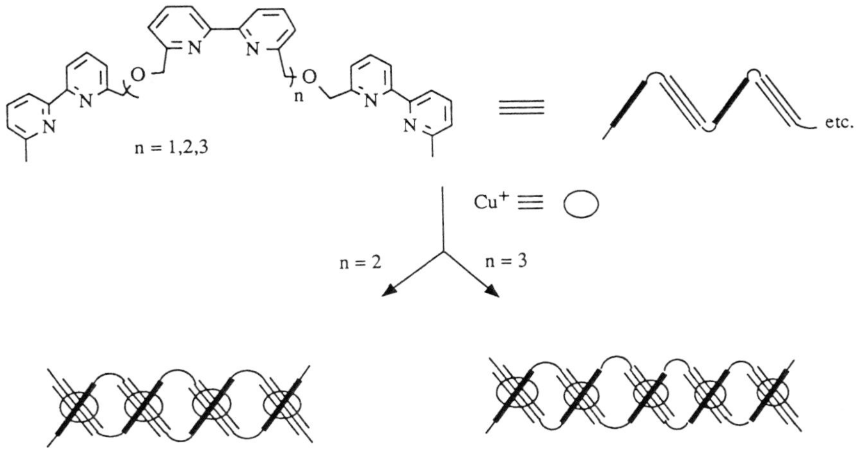

n = 1,2,3

etc.

Cu⁺ ≡ ○

n = 2 n = 3

Scheme 5.29

5.9.2 π-Complexed Catenane Formation

Stoddart and co-workers isolated an interesting complex (**64**) between *bis-para*-phenylene-30-crown-10, a compound originally prepared in Cram's group, and paraquat. Williams and co-workers obtained crystal structure data showing that the complex had 1:1 stoichiometry and that the aromatic rings had their π-systems stacked.

The molecular box in which 4,4′-bipyridyl comprises the sides is electron deficient and can accommodate 1,4-dimethoxybenzene within its cavity (**65a**). The electronic interaction here is similar to that in the complex pictured in **64** except that the guest is electron rich in this case and electron poor in the macrocycle complex.

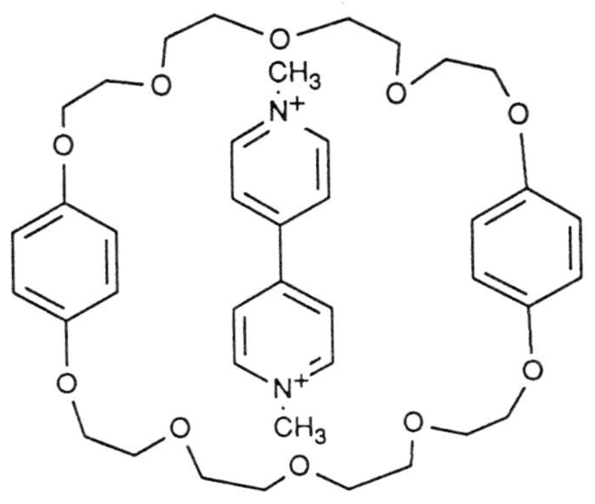

64

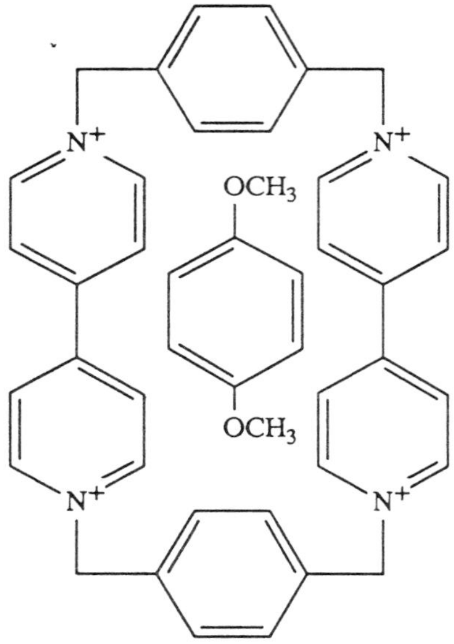

65a

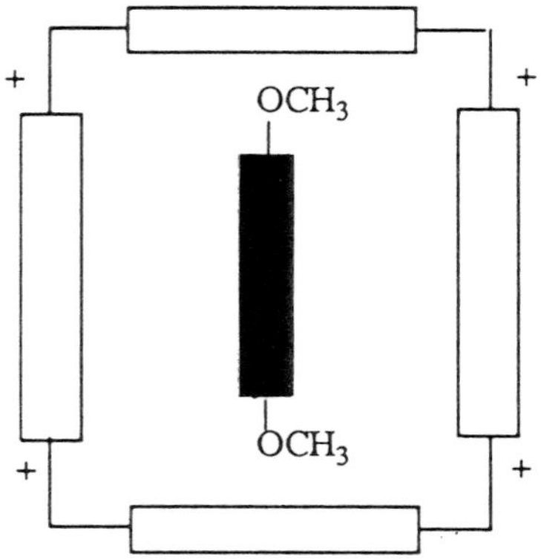

65b

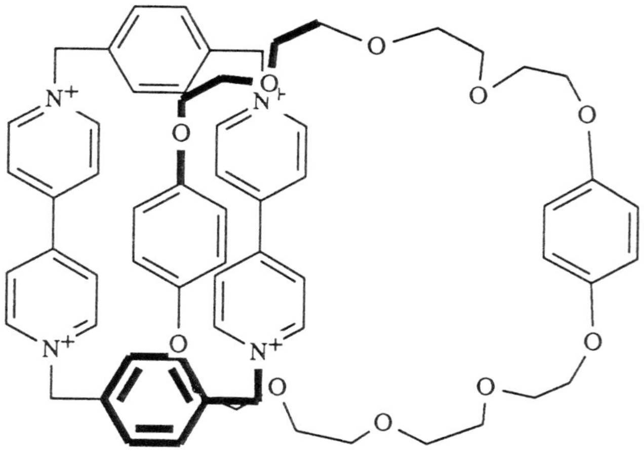

66

Stoddart reasoned that since electron rich π-systems complexed electron poor ones and *vice versa*, that he could interlock the compounds under study. Indeed, he was successful in doing so and formed the complex illustrated in **66**.

The structures now originating from Stoddart's group are so remarkable and complex that he has resorted to the box-diagrams as shown in **65b** of *bis*(viologen). Apparently, one molecule organizes the next and the possibility of forming huge interlocking systems is already being brought to reality.

5.10 The Future of Crown Ether Chemistry

The great diversity of concepts and findings present in this limited volume should suggest to the reader that the future of this area lies not in crown ethers or cryptands, but in the supermolecules that may be formed in part from them. Crowns and cryptands have been remarkably useful in defining new limits to the novel chemical structures that can be conceived and brought to hand. Future work will no doubt focus on specific means to recognize the cations, molecules, and perhaps even supermolecular species.

The future of this area is great because the crown and cryptand studies have dramatically enhanced our field of vision. We now can begin to design structures to have functions hitherto found only in natural products. Indeed, we can now imagine preparing relatively simple (compared to enzymes) structures that can mimic Nature or even better Her. We can now plan the synthesis of materials that are not available at all in Nature and we have the means to produce them. In two decades, the beginning graduate student may have difficulty recognizing the origins of supramolecular chemistry in 18-crown-6 or [2.2.2]-cryptand, but how many of us saw our own remarkable field when we looked at polyethylene glycol molecules three decades ago?

CHAPTER 6

Additional Reading

6.1 Books on Crown Ethers, Cryptands, and Polyethers

J. L. Atwood, J. E. D. Davies, and D. D. MacNicol, 'Inclusion Compounds, Structural Aspects of Inclusion Compounds formed by Inorganic and Organometallic Host Lattices', Academic Press, New York, 1984.

J. L. Atwood, J. E. D. Davies, and D. D. MacNicol, 'Inclusion Compounds, Structural Aspects of Inclusion Compounds formed by Organic Host Lattices', Academic Press, New York, 1984.

J. L. Atwood, J. E. D. Davies, and T. Osa, 'Clathrate Compounds, Molecular Inclusion Phenomena, and Cyclodextrins', D. Reidel Publishing Company, Dordrecht, 1984.

J. L. Atwood and J. E. D. Davies (Eds.), 'Inclusion Phenomena in Inorganic, Organic, and Organometallic Hosts', D. Reidel Publishing Company, Dordrecht, 1987.

F. de Jong and D. N. Reinhoudt, 'Stability and Reactivity of Crown Ether Complexes', Academic Press, New York, 1981.

M. Dobler, 'Ionophores and their Structures', Wiley–Interscience, New York, 1981.

G. W. Gokel and S. H. Korzeniowski, 'Macrocyclic Polyether Chemistry', Springer Verlag, Berlin, 1982.

G. W. Gokel and K. Koga (Eds.), Proceedings of the US–Japan Joint Symposium on Molecular Recognition. 1989.

G. W. Gokel (Ed.) 'Advances in Supramolecular Chemistry', Volume 1, JAI Press, 1990.

M. Hiraoka, 'Crown Compounds: Their Characteristics and Applications', Elsevier, Amsterdam, 1982.

Y. Inoue and G. W. Gokel (Eds.), 'Cation Binding by Macrocycles', Marcel Dekker, New York, 1990.

R. M. Izatt and J. J. Christensen (Eds.), 'Synthetic Multidentate Macrocyclic Compounds', Academic Press, New York, 1978.

R. M. Izatt and J. J. Christensen (Eds.), 'Progress in Macrocyclic Chemistry', Volume 1, Wiley-Interscience, New York, 1978.

R. M. Izatt and J. J. Christensen (Eds.), 'Progress in Macrocyclic Chemistry', Volume 2, Wiley-Interscience, New York, 1981.

R. M. Izatt and J. J. Christensen (Eds.), 'Synthesis of Macrocycles: The Design of Selective Complexing Agents, Progress in Macrocyclic Chemistry', Volume 3, Wiley-Interscience, New York, 1987.

K. Kimura (Ed.), 'Current Topics in Macrocyclic Chemistry in Japan', Hiroshima University School of Medicine Publishing Service, 1987.

G. A. Melson (Ed.), 'Coordination Chemistry of Macrocyclic Compounds', Plenum Press, New York, 1980.

S. Patai (Ed.), 'The Chemistry of the Functional Groups, Supplement E: The Chemistry of Ethers, Crown Ethers, *etc.*', Part 1, Wiley-Interscience, New York, 1981. (New edition, 1989).

Georges Van Binst (Ed.), 'Design and Synthesis of Organic Molecules Based on Molecular Recognition' (Proceedings of the XVIIIth Solvay Conference on Chemistry), Springer Verlag, Berlin, 1986.

F. Vögtle and E. Weber, 'Host-Guest Complex Chemistry: Macrocycles', Springer Verlag, Heidelberg, 1985.

6.2 Monographs on Phase Transfer Catalysis

A. Brandström, 'Preparative Ion Pair Extraction', Apotakarsocieten/Hässle, Läkemedel, Sweden, 1974.

P. Caubere, 'Le Transfer de Phase et Son Utilisation en Chimie Organique', Masson, Paris, 1982.

E. V. Dehmlow and S. S. Dehmlow, 'Phase Transfer Catalysis', 2nd Edn., Verlag Chemie, Weinheim, 1983.

W. E. Keller, 'Phase Transfer Reactions: Fluka Compendium', Georg Thieme Verlag, Stuttgart, Vols. 1 (1986) and 2 (1987).

C. M. Starks and C. Liotta, 'Phase Transfer Catalysis: Principles and Techniques', Academic Press, Inc., New York, 1978.

W. P. Weber and G. W. Gokel, 'Phase Transfer Catalysis in Organic Synthesis', Springer Verlag, Berlin, 1977.

6.3 Reviews and Articles

G. Y. Adachi and Y. Hirashima, 'Macrocyclic Complexes of Lanthanide Ions', in 'Cation Binding by Macrocycles', ed Y. Inoue and G. W. Gokel, Marcel Dekker, New York, 1990, pp. 701–742.

J. L. Atwood, 'Cation Complexation by Calixarenes', in 'Cation Binding by Macrocycles', ed. Y. Inoue and G. W. Gokel, Marcel Dekker, New York, 1990, pp. 581–598.

J. E. Baldwin and P. Perlmutter, 'Bridged, Capped, and Fenced Porphyrins,' *Top. Curr. Chem.*, 1984, **121**, 181.

R. A. Bartsch, 'Complexation of Arenediazaonium Ions by Multidentate Ligands,' *Prog. Macrocyclic Chem.*, 1981, **2**, 1–40.

E. Blasius and K.-P. Janzen, 'Analytical Applications of Crown Compounds and Cryptands,' *Top. Curr. Chem.*, 1981, **98**, 163.

J. S. Bradshaw, 'Synthesis of Multidentate Compounds,' in R. M. Izatt and J. J. Christensen (Eds.) 'Synthetic Multidentate Macrocyclic Compounds', Academic Press, New York, 1978, pp. 53–110.

R. Breslow, 'Biomimetic Control of Chemical Selectivity', in Georges Van Binst (Ed.), 'Design and Synthesis of Organic Molecules Based on Molecular Recognition' (Proceedings of the XVIIIth Solvay Conference on Chemistry), Springer Verlag, Berlin, 1986, pp. 185–197.

R. L. Bruening, R. M. Izatt, and J. S. Bradshaw, 'Understanding Cation-Macrocycle Binding Selectivity in Single-solvent, Extraction, and Liquid Membrane Systems by Quantifying Thermodynamic Interactions', in Y. Inoue and G. W. Gokel (Eds.) 'Cation Binding by Macrocycles', Marcel Dekker, New York, 1990, pp. 111–132.

A. Collet, 'Inclusion Compounds of Cyclotriveratrylene and Related Hosts', in J. L. Atwood, J .E. D. Davies, and D. D. MacNicol (Eds.), 'Inclusion Compounds, Volume 2, Structural Aspects of Inclusion Compounds formed by Organic Host Lattices', Academic Press, London, 1984, pp. 97–122.

H. M. Colquhoun, J. F. Stoddart, and D. J. Williams, 'Second-Sphere Coordination – a Novel Role for Molecular Receptors', *Angew. Chem., Int. Ed. Engl.*, 1986, **25**, 487.

D. J. Cram and K. N. Trueblood, 'Concept, Structure, and Binding in Complexation', *Top. Curr. Chem.*, 1981, **98**, 43.

E. Weber and F. Vögtle, 'Crown Compounds – An Introductory Overview', *Top. Curr. Chem.*, 1981, **98**, 1.

D. J. Cram, 'Preorganization – From Solvents to Spherands', *Angew. Chem., Int. Ed. Engl.*, 1986, **25**, 1039.

D. J. Cram, 'Designed Host–Guest Relationships', in Georges Van Binst (Ed.), 'Design and Synthesis of Organic Molecules Based on Molecular Recognition' (Proceedings of the XVIIIth Solvay Conference on Chemistry), Springer Verlag, Berlin, 1986, pp. 153–172.

D. J. Cram, 'The Design of Molecular Hosts, Guests, and Their Complexes', (1987 Nobel Prize Lecture) *J. Inclusion Phenom.*, 1988, **6**, 397.

D. J. Cram, 'Crystal-structure Windows for Viewing Host–Guest Complexes', *Chemtracts Organic Chem.*, 1988, **1**, 89.

N. K. Dalley, 'Structural Studies of Synthetic Macrocyclic Molecules and their Cation Complexes', in R. M. Izatt and J. J. Christensen (Eds.), 'Synthetic Multidentate Macrocyclic Compounds', Academic Press, New York, 1978, pp. 207–244.

B. Dietrich, 'Cryptate Complexes', in J. L. Atwood, J. E. D. Davies, and D. D. MacNicol (Eds.), 'Inclusion Compounds, Volume 2, Structural Aspects of Inclusion Compounds formed by Organic Host Lattices',

Academic Press, London, 1984, pp. 337–406.

J. L. Dye, 'The Role of Crown and Cryptand Complexation of Cations in the Formation of Metal-Amine and Metal Ether Solutions', *Prog. Macrocyclic Chem.*, 1979, **1**, 63–114.

F. Ebmeyer and F. Vögtle, 'On the Way from Small to Very Large Molecular Cavities', *Bioorg. Chem. Frontiers*, 1990, **1**, 143–160.

E. M. Eyring and S. Petrucci, 'Rates and Mechanisms of Complexation Reactions of Cations with Crown Ethers and Related Macrocycles', in Y. Inoue and G. W. Gokel (Eds.), 'Cation Binding by Macrocycles', Marcel Dekker, New York, 1990, pp. 179–202.

J. H. Fendler, 'Host–Guest Systems', Chapter 7 in 'Membrane Mimetic Chemistry', Wiley–Interscience, New York, 1982, pp. 184–208.

R. B. Fox, 'Nomenclature of Macrocyclic Compounds by Sequential Citation', *J. Chem. Inf. Comput. Sci.*, 1984, **24**, 266–271.

F. R. Fronczek and R. D. Gandour, 'Crystallography of Cation Complexes of Lariat Ethers', in Y. Inoue and G. W. Gokel (Eds.), 'Cation Binding by Macrocycles', Marcel Dekker, New York, 1990, pp. 311–362.

T. M. Fyles, 'Biomimetic Ion Transport with Synthetic Transporters', *Bioorg. Chem. Frontiers*, 1990, **1**, 71–114.

T. M. Fyles, 'Electrostatic Ion Binding by Synthetic Receptors', in Y. Inoue and G. W. Gokel (Eds.), 'Cation Binding by Macrocycles', Marcel Dekker, New York, 1990, pp. 203–252.

G. W. Gokel and H. D. Durst, 'Principles and Synthetic Applications in Crown Ether Chemistry', *Synthesis*, 1976, 168.

G. W. Gokel and H. D. Durst, 'Crown Ether Chemistry: Principles and Applications', *Aldrichimica Acta*, 1976, **9**, 3.

G. W. Gokel, D. J. Cram, C. L. Liotta, H. P. Harris, and F. L. Cook, '18-Crown-6', *Org. Synth.*, 1977, **57**, 30.

G. W. Gokel, D. M. Dishong, R. A. Schultz, and V. J. Gatto, 'Synthesis of Aliphatic Azacrown Compounds', *Synthesis*, 1982, 997.

G. W. Gokel, M. F. Ahern, J. R. Beadle, S. H. Korzeniowski, and A. Leopold, 'Applications of Phase Transfer Catalysis to Arenediazonium Cation Chemistry', *Isr. J. Chem.*, 1985, **26**, 270–276.

G. W. Gokel, D. M. Goli, C. Minganti, and L. Echegoyen, 'Clarification of the Hole-Size Cation Diameter Relationship in Crown Ethers and a New Method for Determining Calcium Cation Homogeneous Equilibrium Binding Constants', *J. Am. Chem. Soc.*, 1985, **105**, 6786.

G. W. Gokel and L. Echegoyen, 'Lariat Ethers in Membranes and As Membranes', *Bioorg. Chem. Frontiers*, 1990, **1**, 115–142.

G. W. Gokel and J. E. Trafton, 'Cation Binding by Lariat Ethers', in Y. Inoue and G. W. Gokel (Eds.), 'Cation Binding by Macrocycles', Marcel Dekker, New York, 1990, pp. 253–310.

I. Goldberg, 'Complexes of Crown Ethers with Molecular Guests', in J. L. Atwood, J. E. D. Davies, and D. D. MacNicol (Eds.), 'Inclusion Compounds, Volume 2, Structural Aspects of Inclusion Compounds formed by Organic Host Lattices', Academic Press, London, 1984, pp. 261–336.

R. Hilgenfeld and W. Saenger, 'Structural Chemistry of Natural and Synthetic Ionophores and their Complexes with Cations', *Top. Curr. Chem.* 1982, **101**, 1–82.

Y. Inoue, Y. Liu, and T. Hakushi, 'Thermodynamics of Cation–Macrocycle Complexation: Enthalpy–Entropy Compensation', in Y. Inoue and G. W. Gokel (Eds.), 'Cation Binding by Macrocycles', Marcel Dekker, New York, 1990, pp. 1–110.

Y. Inoue and T. Hakushi, 'Unsaturated Ethers: Synthesis, Cation Binding, and Reactions', in Y. Inoue and G. W. Gokel (Eds.), 'Cation Binding by Macrocycles', Marcel Dekker, New York, 1990, pp. 523–548.

R. M. Izatt, D. J. Eatough, and J. J. Christensen, 'Thermodynamics of Cation–Macrocycle Compound Interaction', in 'Structure and Bonding, Volume 16, Alkali Metal Complexes with Organic Ligands', Springer Verlag, Berlin, 1973, pp. 161–189.

R. M. Izatt, J. S. Bradshaw, S. A. Nielsen, J. D. Lamb, and J. J. Christensen, 'Thermodynamic and Kinetic Data for Cation–Macrocycle Interaction', *Chem. Rev.*, 1985, **85**, 271–339.

T. A. Kaden, 'Synthesis and Metal Complexes of Azamacrocycles with Pendant Arms having Additional Ligating Groups', *Top. Curr. Chem.*, 1984, **121**, 157.

A. E. Kaifer and L. Echegoyen, 'Redox Control of Cation Binding in Macrocyclic Systems', in Y. Inoue and G. W. Gokel (Eds.), 'Cation Binding by Macrocycles', Marcel Dekker, New York, 1990, pp. 363–396.

R. M. Kellogg, 'Bioorganic Modelling-Stereoselective Reactions with Chiral Neutral Ligand Complexes as Model Systems for Enzyme Catalysis', *Top. Curr. Chem.*, 1982, **101**, 111–145.

K. Kimura and T. Shono, 'Applications of Macrocycles to Ion-Selective Electrodes', in Y. Inoue and G. W. Gokel (Eds.), 'Cation Binding by Macrocycles', Marcel Dekker, New York, 1990, pp. 429–464.

J. D. Lamb, R. M. Izatt, J. J. Christensen, and D. J. Eatough, 'Thermodynamics and Kinetics of Cation-Macrocycle Interaction', Chapter 3 in 'Coordination Chemistry of Macrocyclic Compounds', Plenum Press, New York, 1979, pp. 145–218.

J. D. Lamb, R. M. Izatt, and J. J. Christensen, 'Stability Constants of Cation–Macrocycle Complexes and Their Effect on Facilitated Transport Rates', *Prog. Macrocyclic Chem.*, 1981, **2**, 41–90.

J.-M. Lehn, 'Design of Organic Complexing Agents. Strategies towards Properties', in 'Structure and Bonding, Volume 16, Alkali Metal Complexes with Organic Ligands', Springer Verlag, Berlin, 1973, pp. 1–70.

J.-M. Lehn, 'Molecular Recognition: Design of Abiotic Receptors', in Georges Van Binst (Ed.), 'Design and Synthesis of Organic Molecules Based on Molecular Recognition' (Proceedings of the XVIIIth Solvay Conference on Chemistry), Springer Verlag, Berlin, 1986, pp. 173–184.

J.-M. Lehn, 'Supramolecular Chemistry – Scope and Perspectives: Molecules – Supermolecules – Molecular Devices', (1987 Nobel Prize Lecture) *J. Inclusion Phenom.*, 1988, **6**, 351.

G. W. Liesegang and E. M. Eyring, 'Kinetic Studies of Synthetic Multidentate Macrocyclic Compounds', in R. M. Izatt and J. J. Christensen (Eds.), 'Synthetic Multidentate Macrocyclic Compounds', Academic Press, New York, 1978, pp. 245–288.

S. Lifson, C. E. Felder, A. Shanzer, and J. Libman, 'Biomimetic Macrocyclic Molecules: An Interactived Theoretical–Experimental Approach', in 'Progress in Macrocyclic Chemistry, Volume 3 (Synthesis of Macrocycles – The Design of Selective Complexing Agents)', Wiley, New York, 1987, pp. 241–308.

S. Lindenbaum, J. H. Rytting, and L. A. Sternson, 'Ionophores – Biological Transport Mediators', *Prog. Macrocyclic Chem.*, 1979, **1**, 219–254.

L. F. Lindoy, 'Heavy Metal Chemistry of Mixed Donor Macrocyclic Ligands: Strategies for Obtaining Metal Ion Recognition', in 'Progress in Macrocyclic Chemistry, Volume 3 (Synthesis of Macrocycles – The Design of Selective Complexing Agents)', Wiley, New York, 1987, pp. 53–92.

L. F. Lindoy, 'Mixed Donor Macrocycles: Solution and Structural Aspects of Their Complexation with Transition and Post-Transition Ions', in Y. Inoue and G. W. Gokel (Eds.), 'Cation Binding by Macrocycles', Marcel Dekker, New York, 1990, pp. 599–630.

C. L. Liotta, 'Application of Macrocyclic Polydentate Ligands to Synthetic Transformations', in R. M. Izatt and J. J. Christensen (Eds.), 'Synthetic Multidentate Macrocyclic Compounds', 1978, Academic Press, New York, pp. 111–206.

W. E. Morf, D. Amman, R. Bissig, E. Pretsch, and W. Simon, 'Cation Selectivity of Neutral Macrocyclic and Nonmacrocyclic Complexing Agents in Membranes', *Prog. Macrocyclic Chem.*, 1979, **1**, 1–62.

A. Nakano, Q. Xie, J. V. Mallen, L. Echegoyen, and G. W. Gokel, 'Synthesis of a Membrane-insertable, Sodium-conducting Channel: Kinetic Analysis by Dynamic ^{23}Na NMR', *J. Am. Chem. Soc.*, 1990, **112**, 1287–1289.

J.-Y. Ortholand, A. M. Z. Slawin, N. Spencer, J. F. Stoddart, and D. J. Williams, *Angew. Chem., Int. Ed. Engl.*, 1989, **28**, 1394–1399.

M. Ouchi, T. Hakushi, and Y. Inoue, 'Complexation by Crown Ethers of Low Symmetry', in Y. Inoue and G. W. Gokel (Eds.), 'Cation Binding by Macrocycles', Marcel Dekker, New York, 1990, pp. 549–580.

G. R. Painter and B. C. Pressman, 'Dynamic Aspects of Ionophore Mediated Membrane Transport', *Top. Curr. Chem.*, 1982, **101**, 83–110.

C. J. Pedersen, 'Macrocyclic Polyethers: Dibenzo-18-crown-6 Polyether and Dicyclohexyl-18-crown-6 Polyether', *Org. Synth.*, 1972, **52**, 66.

C. J. Pedersen, 'Synthetic Multidentate Macrocyclic Compounds', in R. M. Izatt and J. J. Christensen (Eds.), 'Synthetic Multidentate Macrocyclic Compounds', Academic Press, New York, 1978, pp. 1–52.

C. J. Pedersen, 'The Discovery of Crown Ethers', (1987 Nobel Prize Lecture) *J. Inclusion Phenom.*, 1988, **6**, 337.

N. S. Poonia, 'Multidentate Macromolecules: Principles of Complexation with Alkali and Alkaline Earth Cations', *Prog. Macrocyclic Chem.*, 1979, **1**, 115–156.

A. I. Popov and J.-M. Lehn, 'Physicochemical Studies of Crown and Cryptate Complexes', Chapter 9 in 'Coordination Chemistry of Macrocyclic Compounds', Plenum Press, New York, 1979, pp. 537–602.

P. G. Potvin and J.-M. Lehn, 'Design of Cation and Anion Receptors, Catalysts, and Carriers', in 'Progress in Macrocyclic Chemistry, Volume 3 (Synthesis of Macrocycles – The Design of Selective Complexing Agents)', Wiley, New York, 1987, pp. 167–240.

D. N. Reinhoudt and F. de Jong, 'Crown Ethers and Related Macrocycles with Bis(methylene)aromatic or -Heteroaromatic Subunits: Their Synthesis and Complexation', *Prog. Macrocyclic Chem.*, 1979, **1**, 157–218.

L. Rossa and F. Vögtle, 'Synthesis of Medio- and Macrocyclic Compounds by High Dilution Principle Techniques', *Top. Curr. Chem.*, 1982, **113**, 1.

R. A. Schwind, T. J. Gilligan, and E. L. Cussler, 'Developing the Commercial Potential of Macrocyclic Molecules', in R. M. Izatt and J. J. Christensen (Eds.), 'Synthetic Multidentate Macrocyclic Compounds', Academic Press, New York, 1978, pp. 289–308.

K. Saigo, N. Kihara, Y. Hashimoto, R.-J. Lin, H. Fujimura, Y. Suzuki, and M. Hasegawa, 'Synthesis and Selective Molecular Recognition of a Macrotricyclic Receptor Having Crown Ether and Cyclophane Subunits as Binding Sites', *J. Am. Chem. Soc.*, 1990, **112**, 1144–1150.

S. Shinkai and O. Manabe, 'Photocontrol of Ion Extraction and Ion Transport by Photofunctional Crown Ethers', *Top. Curr. Chem.*, 1984, **121**, 67.

S. Shinkai, 'Functionalization of Crown Ethers and Calixarenes: New Applications as Ligands, Carriers, and Host Molecules', *Bioorg. Chem. Frontiers*, 1990, **1**, 161–196.

S. Shinkai, 'Dynamic Control of Cation Binding', in Y. Inoue and G. W. Gokel (Eds.), 'Cation Binding by Macrocycles', Marcel Dekker, New York, 1990, pp. 397–428.

W. Simon, W. E. Morf, and P. Ch. Meier, 'Specificity for Alkali and Alkaline Earth Cations of Synthetic and Natural Organic Complexing Agents in Membranes', in 'Structure and Bonding, Volume 16, Alkali Metal Complexes with Organic Ligands', Springer Verlag, Berlin, 1973, pp. 113–160.

J. Smid, 'Solute Binding to Polymers with Macroheterocyclic Ligands', *Prog. Macrocyclic Chem.*, 1981, **2**, 91–172.

J. Smid and R. Sinta, 'Macroheterocyclic Ligands on Polymers', *Top. Curr. Chem.*, 1984, **121**, 105.

J. F. Stoddart, 'Synthetic Chiral Receptor Molecules from Natural Products', *Prog. Macrocyclic Chem.*, 1981, **2**, 173–250.

J. F. Stoddart and R. Zarzycki, 'Second-Sphere Coordination of Transition Metal Complexes by Crown Ethers', in Y. Inoue and G. W. Gokel (Eds.), 'Cation Binding by Macrocycles', Marcel Dekker, New York, 1990, pp. 631–700.

M. Takagi and K. Ueno, 'Crown Compounds as Alkali and Alkaline Earth Metal Ion Selective Chromogenic Reagents', *Top. Curr. Chem.*, 1984, **121**, 39.

M. Takagi, 'Complexation by Chromoionophores', in Y. Inoue and G. W. Gokel (Eds.), 'Cation Binding by Macrocycles', Marcel Dekker, New York, 1990, pp. 465–496.

Y. Takeda, 'The Solvent Extraction of Metal Ions by Crown Compounds', *Top. Curr. Chem.*, 1984, **121**, 1.

Y. Takeda, 'Conductometric Behavior of Cation–Macrocycle Complexes in Solutions', in Y. Inoue and G. W. Gokel (Eds.), 'Cation Binding by Macrocycles', Marcel Dekker, New York, 1990, pp. 133–178.

M. R. Truter, 'Structures of Organic Complexes with Alkali Metal Ions', in 'Structure and Bonding, Volume 16, Alkali Metal Complexes with Organic Ligands', Springer Verlag, Berlin, 1973, pp. 71–112.

H. Tsukube, 'Cation Binding by Natural and Modified Ionophores', in Y. Inoue and G. W. Gokel (Eds.), 'Cation Binding by Macrocycles', Marcel Dekker, New York, 1990, pp. 497–522.

F. Vögtle, H. Sieger, and W. M. Mueller, 'Complexation of Uncharged Molecules and Anions by Crown-type Host Molecules', *Top. Curr. Chem.*, 1981, **98**, 107.

E. Weber, 'Crystalline Uncharged-Molecule Inclusion Compounds of Unintended and Designed Macrocyclic Hosts', in 'Progress in Macrocyclic Chemistry, Volume 3 (Synthesis of Macrocycles – The Design of Selective Complexing Agents)', Wiley, New York, 1987, pp. 337–420.

Subject Index

Acetal, 24
Acylium cations, 89
Adamatane, 64
Allosteric crown, 167
Alkylammonium, 169
Allyl–aluminum. 3
N-Allylaza-18-crown-6, 48
Ammonium ion complexation, 83
Anion activation, 137
Anion complexation, 95
Anthraquinone podand, 156
Arenediazonium cations, 87
Arenediazonium cation,
  complexation of crowns, 87
  coupling of, 89
  entropy effect, 87
Aromatic substitution, 143
12-Ane-4, 44
Anhydrotetramer, 3
Anthraquinone cryptand, 17
Azacrown, 14
Azacrowns, 42
Aza-12-crown-4, sodium complex, 49, 107
Aza-18-crown-6, 134
  bis(nitrobenzyl), 155
  water complex, 123
  potassium complex, 123
Azide anion, 95
Azobenzene, 150
Azobenzene crown, 149
Azocryptand, 155

Back-biting crown, 150
Benzocryptand, 12, 108
Benzo-15-crown-5, sodium complex, 105
Benzo-18-crown-6, 30
BiBLE, 51
Bibracchial lariat ether, complexes, 114
Bicyclic amines, 9
Bicyclic cryptand, 16
Binaphthocrown, 13
Bis(binaphthyl) crown, 164
Bis-crown, 125
Bis(phenol), 5
Binding dynamics, 79
Bistren cryptand, 96
Bola-amphiphile, 160
Bolaform channel, 160
Bolyte, 161
tert-Butylammonium ion complex, 123

Calixarene, 63, 146, 147
Calorimetry, 70
Cannizzarro reaction, 136
Carbanions, 12
Carbonate bases, 144
Carbylamine reaction, 144
Carcerand, 60
Catenane, 175
Catenane formation, 173
Cation binding, *see also* stability constants, 71
  dynamics, 79
  homogeneous solution, 71
  solvent effect, 74
  table, 74, 79
Cation channel, 158
Cation-conducting channels, 160
Cation complexation equilibrium, 72
Cation deactivation, 136
Cation extraction, 66

Cation transport, 81, 158
  pH switching, 146
Cavitand, 16, 60
Charge transfer complex, 176
Chiral barrier, 13, 86
Chiral recognition, 86, 165
Chirality, 164
18-Crown-6, 84
2-Chloroanisole, 133
Chlorophyll, 53
Cholesteryl crown, 163
Chromoionophore, 148
Cholesteryl crown, 100
Chromophore, 147
Colorimetric analysis, 149
Complexation, 64
  ammonium cations, 83
  anion, 95
  azide, 95
  dioxygen, 96
  molecular, 92
  imidazole, 94
  nucleotide, 97
  organic cations, 83
  oxygen, 96
  trihydroxybenzene, 94
  urea, 94
  of urea, 91
Complexation rates, 79
  table, 80
Complexes, neutral 94
Concentric tube, 82
Conductive, 72
Convergent donors, 24
Copper–phenanthroline complex, 173
Coronand, 20
Coronate complexes, 131
Crowns,
  structure of incomplexed, 99
  synthesis, 22
  synthetic variables, 34
12-Crown-4, sodium complex, 106
15-Crown-5, potassium complex, 108
18-Crown-6, 29
18-Crown-6,
  acetonitrile complex, 92
  acetone complex, 93
  benzene sulfonamide complex, 93
  cesium complex, 108
  dimethylthallium complex, 93
  dimethyl acetylenedicarboxylate complex, 9
  hydrazine complex, 122
  hydronium ion complex, 12
  potassium complex, 10
  sodium complex, 109
  structure, 100
  synthesis of, 28
  TCNE complex, 93
  trinitrobenzene complex, 93
27-Crown-9, 84
  ammonium ion complexes, 120
  intramolecular complex, 125
  polymetric, 156
  water complex, 125
Crown activated reagents, 133
Crown esters, 54
  aromatic subunits, 41
  complexation, 64
  pyridine containing, 53
  pyridine in, 53
  self solvating base, 133
  structural variation, 70
  sulfur containing, 56
Crowns, applications, 129
Crown synthesis,
  general conditions, 40
  solvent for, 38
Cryptand, 10
  applications, 129
  complexation, 64
  ditopic, 118
  examples, 76
  exclusion complex, 11
  inclusion complex, 119
  lipophilic, 140
  structure of uncomplexed, 99
  synthesis, 22, 58
[2.2.1]-Cryptand,
  potassium complex, 119
  sodium complex, 116
[2.2.2]-Cryptand, 76
  potassium complex, 116
  structure, 115
  synthesis of, 35
Cryptate, 19
Cryptate complexes, 115
Crystal structure, 99
Cyclam, 27, 44
Cyclic voltammetry, 152

Cyclohexano-15-crown-5, 79
Cyclotetradecane, 64
Cyclotriveratrylene, 63
Cyanosilylation, 141
Cyclodextrin, 167

Diammonium ion complex, 171
Diaza-15-crown-5, 52
Diaza-18-crown-6, 49
  bis(allyl) structure, 103
  bis(azobenzene), 149
  bis(benzyl) structure, 103
  bis(2-hydroxyethyl), 114
  bis(2-methoxyethyl), 114
  structure, 101
Diazomethane, 144
Dibenzo-18-crown-6, 4, 5, 22
  product mixtures, 33
  sodium complex, 109
  synthesis scheme, 6
Dibinaphthyl crown, 164
Dibenzo-30-crown-10, potassium complex 110
Diborane, 47
1, 2-Dichlorobenzene, 133
Dicyclohexano-18-crown-6, 31, 79, 132, 135
  potassium permanganate complex, 132
1, 3-Dihydroxybenzene, 2
Displacement reactions, 14
Disulfide crown, 152
Dithia crowns, 102
Ditopic cryptands, 118
  complexes, 118
Ditopic receptor, 170, 172, 173

EO-4, 3, 21
Electride, 95
Enantiomeric discrimination, 13
Enthalpy, 75
Enthalpy–entropy, compensation, 76
Entropy, 75
Enzyme model, 164
Ester cleavage, 132
Esters, reduction of, 48
Ethylene oxide, 3
  cyclo-oligomerization of, 3, 30
Ethyleneoxy unit, 23
Exclusion complex, 119

Extraction equilibrium, 70
Extraction technique, 65, 66
Extractive alkylation, 137

Ferrocene crown, 152
Ferrocene cryptand, 152
Fluorene indicator, 157
Fluorenyl crown, 165
Furan–acetone tetramer, 3, 13, 29

Gramicidin, 160
Guanidinium complex, 84

Helicate, 173
Hemoglobin, 53
Heterocryptand, 12
High dilution, 27
Hofmann carbylamine reaction, 144
Hole size, 74, 103
Homogeneous stability constants, 73
Hydrazine complex, 122
Hydrogen bonds, 84
Hydronium ion complex, 124
Hydronium cation, 89
Hydronium perchlorate, 90

In–in conformation, 9
In–out bicyclic amines, 9
In–out conformation, 9
Inclusion complex, 119
Interlocking systems, 173
Ionophore, 18
  polymeric, 92
Ion-pair extraction, 137
Ion selective electrode, 65, 73
Ion size, 103
Ionization control of binding, 145
Isocyano crown, 160

Lactams, 46
  formation of, 46
  reduction of, 46
Lariat ether, 16, 49
  bibracchial, 50, 51
  complexes, 112
  nitroaromatic, 153
  redox switchable, 155
  steroidal, 163
Ligand topology, 15
Lithium aluminum hydride, 47

Lüttringhaus' macrocycles, 2

Macrocyclic acetal, 24
Macrocyclic effect, 76
Macrocyclic esters, complexes 111
Membrane, 158
  from crowns, 162
Membrane transport, 146, 153
  diagram, 154
Mercury, 167
Merrifield resin, 157
Mesophase, 164
Mesylate, 37
Metal-templated cantenane formation, 173
Methylenedioxy unit, 23
Micelle, 162
  from crowns, 162
Molecular box, 175
Molecular complexation, 92
Monensin, 9

NAD, 56
Naked ion, 141
Neutral crown complexes, 94
Nitroaromatic crown, 153
Niosomes, 163
Nitrogen,
  incorporation in ring, 49
  in crowns, 42
  protection of, 43
Nitrogen-containing macrocycles, 42
NMR, 72
Nobel prize, 1
Nomenclature, 19
  crown, 7
  cryptand, 11
  table, 21
Nonactin, 18, 19
Nucleophilic aromatic substitution, 133
Nucleotide complexation, 97

Octopus molecule, 128
Organic cation complexation, 13
Out–out conformation, 9
Oxidation–reduction switching, 145
Oxy-Cope rearrangement, 135

Palladium, 167
Palladium complex, 121
Paraquat, 175
pH switching, 145
Phase transfer catalysis, 131, 137
  catalysts, 140
  catalyst comparison, 144
  displacements, 141
  mechanistic diagram, 139
  principles, 138
Photochemical control, 150
Photochemical switching, 145
Phenylenedioxy, 25
π-complexed catene, 175
Picramine crown, 148
Pinene oxidation, 133
Podand, 20, 21
  anthraquinone, 156
Podand complexes, 127
Podando–coronand, 50
Podate, 20, 21
Polymeric crown ether. 156
Polyethylene, glycol, 26
Polymeric ionophore, 92
Porphyrin, 3
Potassium hydroxide in toluene, 132
Potassium permanganate, 133
Pressman cell, 81, 159
Propyleneoxy, 24
Prostaglandin, 143
Protease model, 167
Pyridine crown, 146

Quarternary 'onium salt, 144

Receptor molecule, 169
Redox switching, 145, 151
Resolution, 165
  amino acids, 86
Resorcinol, 2
Rotaxane, 89

Second-sphere complex, 85
Selectivity, 67, 69
Self-solvating base, 134
Sensors, 145
Sexi-pyridine, 55
Sodide, 95
Solubilization of salts, 132
Solubilization phenomena, 130
Solvate, 93

Solvation shell, 131
Solvents, 38
  table of, 39
Spherand, 16, 60, 167, 170
  synthesis of, 62
Stability constants, *see also* cation binding, 71
  by calorimetry, 72
  by conductance, 72
  by ion selective electrodes, 73
  by NMR, 72
  techniques, 71
Steroid crown, 162
Stilbene, 150
Supermolecule, 97
Superoxide, 142
Switching modes, 145

Template effect, 6, 27
  principle of, 28
Tetraaza-12-crown-4, 146
Tetracarboxy-18-crown-6, hydronium complex, 124
Tetrahedral receptor, 173
Thermal switching, 145, 151
Thermodynamics, 75
Thermodynamic data, 76, 78
Thiacrown, 14, 56
Thiacryptand, 121
Titration calorimetry, 72
Tosylate, 36
Toxicity, of crowns, 21
Transport,
  cation, 81
  concentrate tube, 82
  U-tube, 81
Triaza-18-crown-6, 52
  structure, 101
Trimethylsilyl cyanide, 141
Tripodand, 128
Tris(binaphthyl) crown, 164
Tropone, 143
Tungsten, 167

U-tube, 81, 159
Uranyl cation, 91

Valinomycin, 18, 158
van't Hoff, 75
Vesicles, 163
Viologen crown, 176

Williamson reaction, 34